MY POWERFUL HOME
PLUG INTO YOUR HOME'S FULL POTENTIAL

Keith Johnson and Janet McConnell Johnson

Published by: Financial Peace Marketing

My Powerful Home

Printed in the Untied States of America.

ISBN 978-0-9845260-0-0

For our children and the empowering of
them with a brighter future.(Pun intended)

My Powerful Home

Contents

About the Authors & Acknowledgment _____ vii

Introduction_____ viii

SECTION ONE
The First Step – Making Your Home Efficient
    Know Your Home_____ ix

    Home Envelope
        Your Existing Home_____ 2
          Air Leaks_____ 2
          Stopping the Chimney Effect_____ 4
          Insulate_____ 4
            Basement_____ 5
            Walls_____ 6
            Attic_____ 6
          Upgrade Inefficient Doors & Windows_____ 8

        Your New Construction
          Insulated Concrete Forms (ICF)_____ 12
          Structural Insulated Panel System (SIPS)_____ 15

    Lighting
        Incandescent_____ 18
        Compact Fluorescent (CFL)_____ 19
        LED_____ 20
        Lighting Comparison_____ 21
        Tubular Daylighting Devices (TDD)_____ 24

    Appliances - Energy Star Savings_____ 27

    Your Home's Systems_____ 31
        Space Heating & Cooling (HVAC)_____ 32
        Ground Source Heating & Cooling
          What is Geothermal Heating & Cooling?_____ 33
          Ground Heat Exchanger_____ 34
            Closed Loop_____ 35
            Open Loop_____ 36
         Types of Ground Source Heat Pumps (GSHP)

Water to Air		37
Water to Water		39
Combination units		40
Split System		41
Options with GSHPs		
ECM		42
Two-stage Compressors		42
Desuperheater		43
Hybrid Heating & Cooling		45
Hot Water Production		47
Desuperheater		47
Solar Hot Water		48
How It Works		48
Hot Water Storage		49
Active Solar Water Heaters		51
Heat Pump Water Heater		53
Instantaneous Water Heater		54
Having an Integrated System		56
Diagram with Explanation		57

SECTION TWO
Next Step - Generating Electricity

Powerful Choices, Empowering Solutions		59
Renewable Energy Systems Diagram		60

Electricity 101

Alternating Current (AC)		62
Transformers		63
Direct Current (DC)		63
Inverters		64
Understanding Volts, Amps & Watts		65
Storing Electricity		69
Types of Batteries		71
Battery Size		73
Battery Life		74
Battery Charging		75
Net Metering		76
Solar Power		81

Photovoltaic (PV) Basics_____ 83
Types of Solar Panels_____ 85
Optimizing the Rays_____ 86
Off the Grid with Solar_____ 88

Wind Power
Wind Basics_____ 92
Wind Generator Sizing_____ 96
Choosing a Wind Turbine_____ 98
Wind Turbine Tower & Height_____ 99

Micro-hydro Power
Intro to Micro-hydro_____ 104
Measuring Head Pressure_____ 106
Measuring Flow_____ 108
Determining Available Power_____ 108
Components of a Micro-hydro System_____ 109
Grid Tie vs. Off Grid_____ 112

Emergency & Power Assist Generators_____ 115
Generator Sizing_____ 117
Generator System Costs_____ 120
Manual Generator System_____ 121
Automatic Standby Generator System_____ 124
Generator Fuel Types_____ 125
AC & DC Generators_____ 126

Tying It All Together_____ 129
Diagram with Explanation_____ 130

The Right Design Saves You Money_____ 133
A Professional Design_____ 136

PLUS
System Costs & Tax Incentives_____ 139

Glossary_____ 148

Index_____ 161

Resources_____ 167

About the Authors

Keith Johnson is a licensed master electrician, journeyman H.V.A.C contractor, I.G.S.H.P.A. certified in ground source heating and cooling, and has more than 30 years experience in residential, commercial, and industrial construction. Keith has also applied his knowledge to photovoltaic, wind, and hydropower generation. It is rare to find someone in the industry that has designed and installed all of these systems.

Janet Johnson is the perfect complement to Keith. Prior to their meeting, Janet was the general contractor for her large mountain house. With her Dad's council, Janet installed the electrical, finish plumbing and radiant floor heating and coordinated all of the trades to complete this very nice home. She is a true 'learn how, then do-it-yourself' type of person.

Special Acknowledgements

We would like to thank Kay General Contracting, KayGeneral.com, for their help and support in putting this book together - Clark Kay with over 35 years experience in the construction industry, LEED accredited professional, IGSHPA certified and NAHB certified Green Professional - Steve Decker with over 25 years experience, NAHB certified Green Professional and IGSHPA certified. In addition, special thanks to SueAnne Pace for her help with the glossary and additional research.

We would like to thank our parents, Keith and Leona Johnson, and Warren and Bonnie McConnell, for all of their help and support over the years, providing the way for us to learn and grow in the construction industry and life in general.

MY POWERFUL HOME

Introduction

Congratulations! You are about to read a book that will truly guide you to make the best choices for home efficiency and renewable energy. An energy-efficient home translates into a cost-effective approach to purchasing and installing your renewable energy systems. When planning a renewable energy system, the first step is to know the current energy requirement of your home and then reduce it through energy efficiency. With lower electricity consumption, you can purchase a smaller solar array, wind turbine, back-up generator, or fewer batteries. This approach ultimately means a financial savings for you.

We have divided this book into two major sections:

The First Step - Making Your Home Efficient - Knowing and Reducing Your Home's Energy Requirements

The Second Step - Generating Electricity - Renewable Equipment & Systems, On or Off the power Grid

Plus:

Costs & Tax Incentives – What is the Bottom Line?

We use many years of practical experience to help you wisely choose the proper systems for your circumstance. Additionally, our informative website, MyPowerfulHome.com is available for your benefit with videos and articles on all of the topics of this book. There, you can join in on the 'Ask Keith' video answers to your energy questions or have your home evaluated and designed for a renewable energy system. The goal of this book and the website is to empower you (pun intended) with what you need to be partially or completely off the power grid. **Knowledge truly is power in the energy market.**

THE FIRST STEP - Making Your Home Efficient

KNOW YOUR HOME

The first step into the world of home energy production is knowing more about the home in which you live or will be building. When you know where energy is lost through the exterior structure or home envelope, then you can begin tightening it up and reduce energy loss. In addition, when you know where and how efficiently most of your electricity is being consumed, you have a clue as to where to make the biggest reduction.

Take a look at this chart, put out by the Department of Energy (DOE), of residential energy use for 2005. It gives you an idea of where to begin. Heating and cooling is 43% of the pie. You don't have to adjust the thermostat to use less energy; try installing an efficient Energy Star heating and cooling system. This will make the biggest energy reduction. Even changing the types of light bulbs in your home

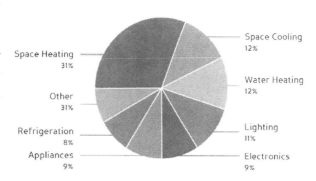

Figure 1.1 Department of Energy residential energy use.

will reduce the amount of electricity used. As you go through this section, remember that reducing energy loss and electricity usage means less energy that you will need to produce. This translates into a smaller renewable energy system to purchase.

In this First Step section, we will address energy loss and energy efficiency measures in these chapters:

The Home Envelope	Existing Home & New Construction: Tightening It Up
Lighting	"Watt" Difference Does a Light Bulb Make?
Appliances	Are All Appliances Created Equal?
Your Home's Systems	Space Heating & Cooling and Hot Water Production

HOME ENVELOPE
Existing Home - New Construction
Tightening It Up

HOME ENVELOPE - YOUR EXISTING HOME

Your Questions:

- **What is the Home Envelope?**
- **Where do I find air leaks in my home?**
- **What is the Chimney Effect and how can I stop it?**
- **How should a home be insulated, and how can I add more?**
- **What can I do to my windows and doors to have a tighter home envelope?**

The amount of energy your home demands is largely affected by the efficiency of the home envelope. The home envelope is the insulated structure that encloses your living space. Problems with the home envelope, such as having unseen air leaks, insufficient insulation, and low quality doors and windows translate into higher energy consumption. In this chapter, we will answer the questions about how to identify and fix home envelope issues. These problems can often be fixed simply and inexpensively.

LEAKS

In the winter, more than any other time of the year, you will notice your home's air leaks. Most people call these air leaks "drafts." You may feel these drafts around windows and doors and think these leaks are your major source of wasted energy. In most homes, however, the most significant air leaks are hidden in the attic and basement. These are the leaks that significantly raise your energy bill and make the temperature in your home uncomfortable.

Common locations where Household Air Leaks
- Behind knee walls
- Plumbing penetrations through insulated floors and ceilings
- Chimney penetrations through insulated ceilings and exterior walls
- Fireplace inserts and dampers
- Attic access hatches
- Wiring penetrations though insulated floors, ceilings and walls

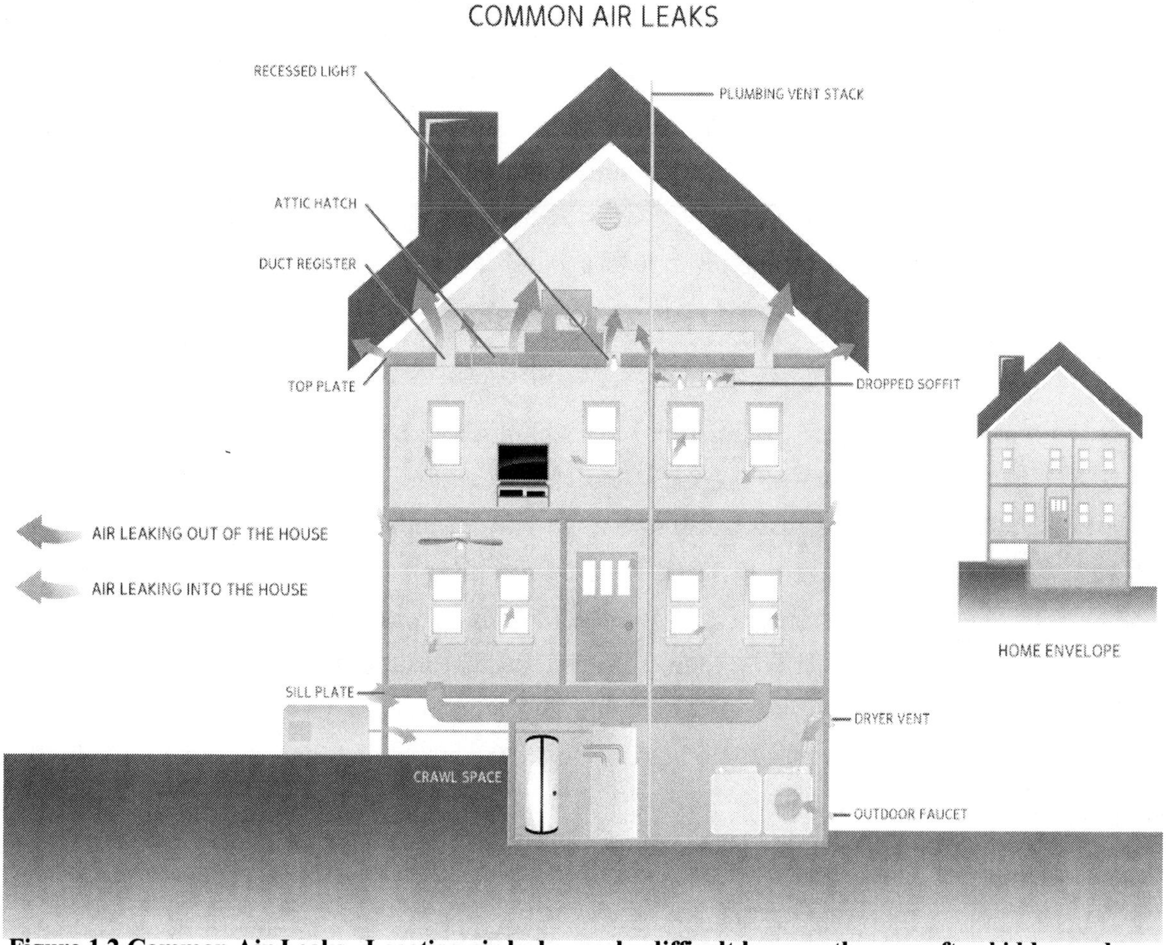

COMMON AIR LEAKS

RECESSED LIGHT

PLUMBING VENT STACK

ATTIC HATCH

DUCT REGISTER

TOP PLATE

DROPPED SOFFIT

AIR LEAKING OUT OF THE HOUSE

AIR LEAKING INTO THE HOUSE

HOME ENVELOPE

SILL PLATE

DRYER VENT

CRAWL SPACE

OUTDOOR FAUCET

Figure 1.2 Common Air Leaks   Locating air leaks can be difficult because they are often hidden under the insulation. In cold weather, warm air rises in your house, just as it does in a chimney. As this air rises up into your attic, it sucks cold air in through every part of your home.

- Electrical outlets and switches, especially on exterior walls
- Window, door and baseboard moldings
- Under bathtubs and showers
- Dropped ceilings above bathtubs and cabinets
- Open soffit
- Recessed lights and fans in insulated ceilings
- Furnace flue or duct chaseway (the hollow box that hides ducts)
- Foundation sill plates (where the foundation meets the wood framing)
- Windows and doors

3

STOPPING THE CHIMNEY EFFECT

In cold weather, warm indoor air rises throughout the home, just like a chimney, and sucks with it cold outdoor air through every crack and hole. This is called the chimney effect. Leaks in the attic increase the amount of outside air drawn in through basement cracks and holes. It is recommended that you seal openings that go through the basement ceiling to the floor above. Generally, these are holes for wires, water supply pipes, water drain pipes, plumbing vent pipes (for venting sewer gases), and furnace flues (for venting furnace exhaust).

Figure 1.3 Use caulking to reduce air leaks

The best material for sealing these hidden air leaks depends on the size of the gaps and their location. Caulk is best for cracks and gaps less than about 1/4" wide. Expanding foam sealant is an excellent material to use for sealing larger cracks and holes that are protected from sunlight and moisture. Backer rod or crack filler is a flexible foam material, usually round in cross-section (1/4" to 1" in diameter), and sold in long coils. Use it to seal large cracks and to provide a backing in very deep cracks that are to be sealed with caulk.

Use rigid foam insulation for sealing very large openings such as plumbing chases and attic hatch covers. Fiberglass insulation can also be used for sealing large holes, but it will work better if wrapped in plastic or stuffed in plastic bags, because air can leak through exposed fiberglass. You can also purchase pre-cut foam sealers to go behind your exterior wall electrical cover plates. Specialized materials such as metal flashing and high-temperature silicone sealants may be required for sealing around chimneys and flue pipes. Check with your building inspector or fire marshal if you are unsure about fire-safe details in these locations.

INSULATE

Insulation is your primary defense against heat loss and air infiltration through the home envelope. Insulation is rated in terms of R-value or the resistance to heat transfer. The higher the R-value, the less heat transfer there is through a particular material. Insulations have high R-values, while materials like wood and concrete have very low R-values. This is why insulation is critical to your home envelope performance. Additionally, insulation is one of the most cost effective up-grades you can perform on your home. Local utility, state and federal insulation tax credits and rebates of up to $1500 makes it more affordable to tighten up your home. (See the chapter on Tax Incentives for more information.) There are some great "How to Install Insulation" videos on the internet to help the 'Do-it-yourself' homeowner. Ownes Corning is one of the companies that has very helpful Youtube videos. Your local hardware store where you purchase insulation can also be a great resource.

Figure 1.4 Various types of insulation for basements, walls and attics.

Let's look at insulating the *basement, walls and attic* of your home.

- *Basement* - Materials that could be damaged by moisture, such as fiberglass batts or cellulose, are not always the best solution to insulate a basement. A better choice for basement insulation is rigid foam installed against the concrete walls. If you are considering finishing your basement and using it as a living space, seek the advice of an experienced building contractor. If you have a crawl space, it should be sealed, not ventilated. To do this, use 6-mm thick polyethylene sheeting as a moisture barrier to cover the ground and seal tightly to walls and columns. Then, use rigid foam to insulate the foundation walls.

- *Walls* - Putting insulation into wall cavities of an existing house that has no insulation can be difficult. In this case, the best option is to bring in an insulation contractor to blow cellulose or fiberglass into the wall cavities.

In new construction, wall cavities in a wood-framed house can be filled with fiberglass batts, blown in cellulose or fiberglass, or spray foam. Spray foam costs quite a bit more, but it has the highest insulating values per inch and

Figure 1.5 Air barrier building wrap highly reduces wall air infiltration.

is extremely effective at sealing up all the gaps in wood-framed construction. In addition to wall insulation, an air barrier building wrap should be installed on the exterior. Since heat is transferred easily through air, the less air infiltration in and out, the better.

- *Attic* - Adding insulation to an unheated attic is easier than insulating existing walls and is likely to have a greater impact on comfort and energy use. If there is no floor in the attic, simply add more insulation between the ceiling joists or trusses; use either loose fill or un-faced fiberglass batts. In most of the U.S., a full foot of fiberglass or cellulose insulation in the ceiling joists is cost-effective. However, it is critical to install fiberglass batts properly in order for them to do the job. You can rent equipment from your local hardware/home repair store to blow in insulation yourself, but make sure you read up on correct installation practices.

The following pictures show another cost-effective energy solution for the attic. This can be achieved by installing a reflective barrier above the rafters prior to laying the wood sheeting or over your existing insulation to reflect the sun's energy away from the attic space in the summer and retain radiant heat inside the house in the winter. With this method, an energy savings of 17% can be achieved according to an Energy Star study (http://www.radiantbarrier.com/energy-savings.htm)

Figure 1.6 Reflective barriers in the attic above insulation or above the trusses reflect radiant heat.

We also need to address another problem in the attic space and give you an appropriate solution. Recessed light fixtures installed all over your ceiling can provide a path for warm air in your home to migrate through the can into the attic and then outside (This is the chimney effect that we discussed earlier). Recessed lights have enclosed metal cans that trap heat very well inside the can; in fact, you can cook an egg on top of many recessed cans! With non-insulated rated cans, codes require that your insulation be blocked from getting too close to your cans. As a result, you have less insulation keeping warm or cool air in your home.

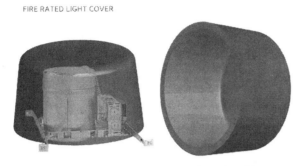

FIRE RATED LIGHT COVER

Figure 1.7 Fire Rated light covers reduce the chimney effect caused by leaks in can lighting.

It is quite revealing to climb into an attic in the wintertime and feel all the heat escaping around these light fixtures. You might as well just cut a large hole in your ceiling or leave a window open all winter long. In the summer, the heat in your attic can reach temperatures up to 160°F, and actually lessen the lifetime of the shingles. A solution for air migration and lack of insulation around recessed light fixtures is to install a protective cover around them like the one we are showing in this drawing.

Now that we have shown you why and where to insulate your home for a higher performing home envelope, let's take a look at how to make windows and doors more energy-efficient.

WINDOWS AND DOORS

About one-third of a home's total heat loss usually occurs through windows and doors. Increasing the efficiency of windows and doors can be accomplished by boosting their efficiency or replacing them.

Boosting efficiency - If your windows are generally in good shape, it will probably be more cost-effective to boost their efficiency with inexpensive products that can be purchased from your local building supply or hardware store. Here are six quick fix tips:

1. Seal all window edges and cracks with caulk or rope caulk. This is the quickest and least expensive option.

2. Weather-strip windows and doors with a special lining that is inserted between the window and the frame. For doors, weather-strip around the top and sides to ensure a tight seal when closed.

3. Install quality door sweeps on the bottom of the doors if they aren't already in place.

4. Plastic films can be attached to the inside of existing windows to increase their efficiency.

5. Install insulating curtains or drapes on the interior of windows and glass doors. This prevents radiant heat gain and loss through the seasons.

6. Install storm windows if you plan to stay in the house for more than a few years. These come as removable pieces of glass, in a frame, attached to the outside of the existing window.

Replacing windows and doors - If your existing windows have rotted or damaged wood, cracked glass, single pane glass, missing putty, poorly fitting sashes, or locks that don't work, you may be better off replacing them. Whether replacing windows in an older house or choosing windows for a new house, your decisions on what type of windows to buy will be among the most important decisions you will make in terms of energy use.

Figure 1.8
Energy Star label

To help consumers choose wisely, the federal government implemented the Energy Star program a few years ago. It was intended to give consumers a way to evaluate the energy efficiency of various building products. When a product, such as a window, appliance, or lighting fixture bares the Energy Star label, this is a signal to the consumer that it meets a high standard of energy efficiency. The labels can be large and highly visible or may be small and hard to find but they are there if the appliances meet the Energy Star standard. Additionally, consumers are rewarded by the government with tax incentives when they install Energy Star products in their homes. We encourage you to search the EnergyStar.org website for details.

Because of the impact windows have on both heat loss and heat gain, proper selection of products can be confusing. Energy-efficient options in windows include frame material, double and triple pane glass, Low-E coatings on glass to reflect UV rays, and gas between glass panels.

For the most up-to-date information on what to look for in energy-efficient windows for your home, visit the Efficient Windows Collaborative at www.efficientwindows.org. Efficient Windows Collaborative (EWC) members have made a commitment to manufacture and promote energy-efficient windows. This site provides unbiased information on the benefits of energy-efficient windows, descriptions of how they work, and recommendations for their selection and use.

This map, created by the Department of Energy, shows the annual savings when you upgrade to Energy Star windows in your home. It is broken down into regions.

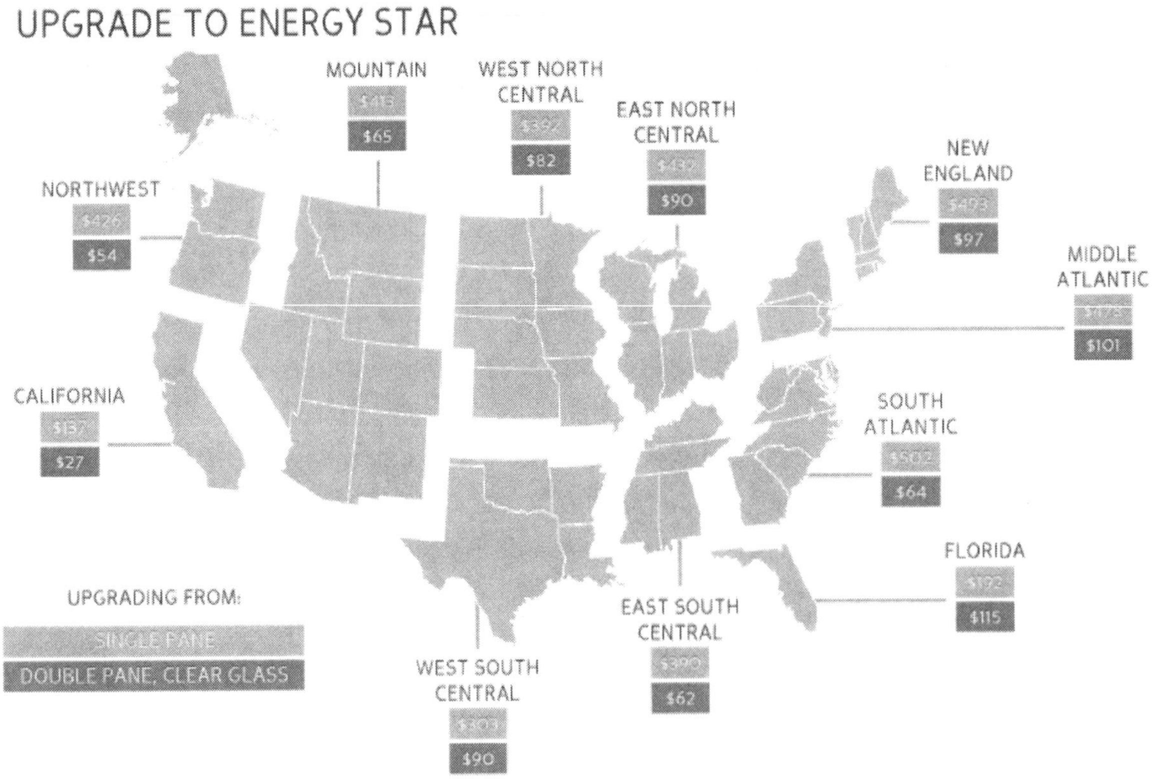

UPGRADE TO ENERGY STAR

MOUNTAIN	$45	$65
WEST NORTH CENTRAL	$392	$82
EAST NORTH CENTRAL	$432	$90
NEW ENGLAND	$493	$97
NORTHWEST	$429	$54
MIDDLE ATLANTIC	$475	$101
CALIFORNIA	$137	$27
SOUTH ATLANTIC	$502	$64
FLORIDA	$192	$115
EAST SOUTH CENTRAL	$390	$62
WEST SOUTH CENTRAL	$303	$90

UPGRADING FROM:
SINGLE PANE
DOUBLE PANE, CLEAR GLASS

Figure 1.9 U.S. Department of Energy savings estimates are based on population-weighted regional annual energy use for a 2,000 sq.ft, single-story, detached house with 300 sq. ft. of window area, gas heat, electric air condition. Estimated by Energy Information Admin's average gas and electricity prices from October 2007 through September 2008. Double pane, clear glass may not be applicable to all areas due to local building codes. Savings vary by climate region and home characteristics.

Important Points:

- **Discovering air leaks and stopping the chimney effect will enhance your comfort and save on utility bills.**

- **Take advantage of tax incentives for adding insulation and replacing windows and doors.**

- **A sealed home envelope with quality doors and windows saves you money when purchasing renewable energy systems.**

HOME ENVELOPE – NEW CONSTRUCTION

Your Questions:

- **What building materials or systems could I utilize to have a superior structure?**

- **Will these systems give me a structure that can withstand extremely high winds and possibly earthquakes?**

Your new home should be constructed with all the energy-efficient techniques we have already discussed and with the best possible walls and roof. When building a new home, the goal should be to build a structure that is superior in strength and efficiency. Building with *Insulated Concrete Forms (ICF)* and *Structural Insulated Panel System (SIPS),* two of our favorite building systems, accomplishes this aim.

ICF

Building residential and commercial buildings with Insulated Concrete Forms (ICFs) has many advantages. They result in a stronger, more efficient, nearly soundproof, and faster built structure, as opposed to traditional building methods such as wood, concrete block, and steel frame construction. To see an example of this type of building process go to My-PowerfulHome.com

Figure 1.10 Insulated Concrete Forms

The ICF form consists of two stay-in-place panels of Expanded Polystyrene (EPS) foam connected with a hard plastic web tie every 6" to 8". The forms are stacked, reinforced, and filled with structural concrete to form a solid, reinforced monolithic concrete wall that can be 4" to 12" thick. They are 16" to 18" high and 4' to 8' long. Preformed 90s, 45s, T-blocks, brick ledge forms and curved forms are available. The pictures of ICF blocks are from Nudura Corp., one of many manufactures.

The ICF blocks consist of six wall components in one:

1. Concrete forming system
2. Wall structure
3. Insulation
4. Air barrier
5. Vapor barrier
6. Backing for interior and exterior finish systems

Having these components in one system, cuts out several construction activities, enabling you to build faster and more efficiently.

The EPS panels are 2½ to 2⅝ inches thick. Coupled with the thermal mass of the concrete, this wall system has an insulating performance of about an R-32 compared to the average R-19 of a wood-framed 2'x6' wall. The blocks are stacked together like Lego blocks, forming an air barrier that provides almost zero air infiltration. These two factors combined contribute to a wall system that is up to two times more energy-efficient than traditional construction. This enables the customer to downsize heating and air conditioning equipment. Coupled with a ground source heat pump (that will be discussed later on), the savings can be incredible.

This photo shows an ICF wall portion in the building process. It is the first row in a wall that will eventually be nineteen feet high. Notice that it has a brick ledge that will support the exterior masonry. With the insulated forms and the required rebar, this building will be strong and well insulated.

Figure 1.11 ICF wall building process.

The second picture shows the use of a concrete pump truck to pour the wall. Braces are used to ensure a straight wall.

Figure 1.12 ICF wall pouring, using a concrete pump truck.

There are five major benefits to building with ICF:

- Strength - The reinforced concrete wall construction provides unmatched strength and is especially suited for areas where earthquakes, hurricanes, and tornados occur. These walls can withstand winds up to 250 MPH!

- Fire - In addition to strength is fire protection. These wall systems provide a 3-4 hour fire protection rating.

- Sound - The EPS panels and concrete provide an excellent sound barrier. Many builders have used them in movie theater construction to prevent sound transfer between multiple theaters.

- Value - Building with ICF is an incredible value for the money. It is close in material cost to a well-built, stick frame construction home. The savings occur in the labor costs of framing, insulating, and installing air and vapor barriers.

- Heating and Cooling – In the short term, the heating and cooling system can be downsized, costing less. In the long term, there are savings in heating & cooling costs because it provides a tighter envelope with less air infiltration and a higher insulation R-Value.

We believe that ICF forms are optimal for exterior walls. The up-front expense is about that of a well-built 2"x 6" building, but with superior strength, fire protection, sound absorption, and air infiltration values.

SIPS Panels

The Structural Insulated Panel System (SIPS) is another non-conventional building method with great strength, rapid construction, and potential energy savings. SIPs combine structure and insulation in one large building panel up to 8'x 24'. For structure, SIPs operate on the same principle as a steel I-beam. They consist of two rigid sheets of Oriented Strand Board (OSB), which act as flanges bonded to a rigid plastic foam core made of Expanded Polystyrene (EPS) foam, which acts as the web.

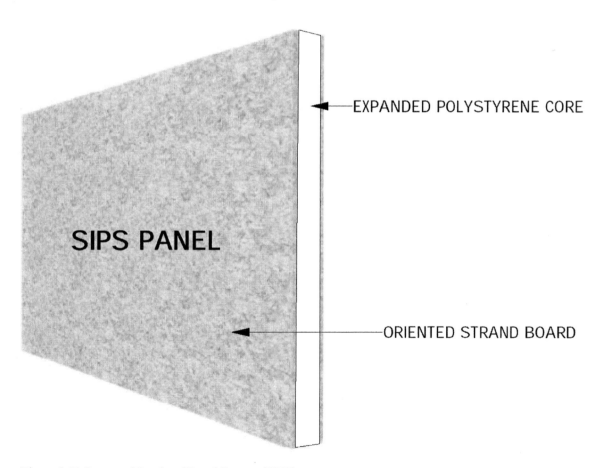

Figure 1.13 Structural Insulated Panel System (SIPS)

These panels can be used in both wall and roof structure. They come in 4½, 6½, 8¼, 10¼, and 12¼-inch thicknesses. The thickness will determine R-value and structural strength per application. Depending on the span of a roof section, trusses or other support may be unnecessary.

For more information on SIPs Panels, visit the Structural Insulations Panel Association (SIPA) website at www.sipa.org. Be sure to watch the great time lapsed video building a SIPS home. You can find it under the *Publication* tab then choose *Videos.*

After showing you both the ICF and SIPS construction options, we suggest building the basement and all exterior walls with ICF and the roof with SIPS. This will create a highly efficient home that is built to weather any storm and last for generations.

Important Points:

- **A structure built with ICF and SIPS will withstand extremely harsh conditions, giving you more peace of mind.**

- **ICF and SIPS are superior in strength, and can also be superior in efficiency with high R-values and zero air infiltration, saving you money on utilities and renewable energy systems.**
- **These structures have a high fire rating and are less likely to sustain heavy damage over their lifetime, saving you money on insurance.**

LIGHTING
"Watt" Difference Does a Light Bulb Make?

LIGHTING

Your Questions:

- **If one light bulb uses the same amount of electricity (watts) as another, will it give off the same amount of light?**
- **Can I overcome the initial cost of new high efficient light bulbs?**
- **How will efficient lighting save money on renewable installations?**
- **How long does high efficient lighting last?**

Understanding how to use different light sources in your home effectively is very important, especially with the goal of reducing your energy consumption. When comparing light bulbs, some last longer, emit equal light, radiate less heat, and use less energy (watts), saving you money. Natural sunlight is free and should also be considered. We will present

Figure 1.14 Compact Fluorescent (CFL), LED and Incandescent light bulbs choices.

lighting information on 1. *Incandescent lights,* 2. *Compact Fluorescent Lighting (CFL),* 3. *LED lights,* 4. *Lighting Comparison, and* 5. *Tubular Daylighting Devices (TDD).*

Incandescent Lights

Many people still use the standard incandescent bulb in their homes because it is the cheapest on the market in terms of cents per bulb. There are many different types of incandescent lighting, such as quartz, halogen, track lighting, floodlights, etc. Although it can come in many forms, it is still the original technology invented by Edison in 1879. They have filaments that glow to produce light, but they also produce tremendous heat. Most people do not take this into consideration when screwing a new bulb into the socket.

Incandescent light bulbs waist about 90% of their power in heat and the remaining 10% is converted to light. So, if you want an inefficient *heat source*, screw in a bunch of incandescent light bulbs and install a larger air conditioner to overcome the heat gain in the summer. These light bulbs generally have a life expectancy of 1000 to 2000 hours before the element burns out. This

Figure 1.15 Incandescent bulbs are very inefficient.

means you will have to climb a ladder to replace those incandescent bulbs in the ceiling 10-50 times more often than with a more efficient type of lighting, discussed next.

Figure 1.16 Compact Fluorescent bulb

Compact Fluorescent Lighting

Compact fluorescent lighting, or CFL, light bulbs require an electronic ballast to boost the voltage and ignite argon and mercury vapor creating light. This type of light generates far less heat, but produces the same amount of light as incandescent lights do. This type of lighting is definitely an improvement in energy efficiency, but not all CFL bulbs are of the same quality.

Some CFL light bulbs have inferior electronics. They can have high quantities of mercury and can fail sooner than their standard 10,000-hour life rating. However, when you buy an Energy Star CFL bulb, you can expect quality electronics and mercury as low as 1 mg per bulb. Pay attention to what you are buy-

ing and start looking for that Energy Star label. It will pay off when you have to replace bulbs less often and you see your electricity bill reduced.

One last tip on CFL lighting: Installing CFLs in enclosed fixtures can create too much heat and shorten their life. Keep these light bulbs protected from outside moisture and pay attention to their temperature rating for cold climates.

Green Note: According to the Energy Star website, if every new home that is built in the United States merely used ENERGY STAR CFL, we would prevent greenhouse gases equivalent to the emissions of nearly 320,000 cars and save more than $250 million/year—the equivalent of paying off the electricity bill for nearly 240,000 homes for a full year.

LED Lights

LED, or light-emitting diode, lighting starts with a tiny chip with layers of semi-conducting material. LED lighting may contain just one chip or multiple chips, mounted on heat-conducting material called a heat sink and usually enclosed in a lens.

LED devices are mounted on a circuit board, which can be programmed to include lighting controls such as dimming, light sensing, and pre-set timing. The circuit board is mounted on another heat sink to manage the heat from all the LEDs in the array. Make sure you

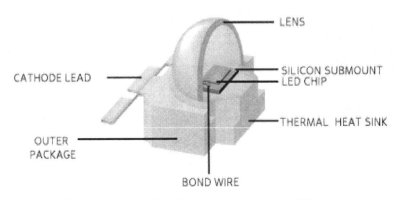

Figure 1.17 Components of an light-emitting diode (LED).

have a compatible dimmer switch if you want to dim your lights. Fixture manufacturers can provide a list of approved dimmer switches.

LED lighting is more efficient, durable, versatile, and long-lasting than incandescent and fluorescent lighting. LEDs emit light in a specific direction, whereas an incandescent or fluorescent bulb emits light heat in all directions. LED lighting emits both light and heat more efficiently. Many manufactures boast that LED lighting can last more than 50k hours.

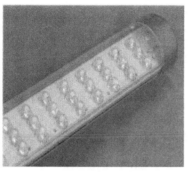

Figure 1.18 Various types of LED bulbs.

Lighting Comparison

With a general understanding of the three major lighting types commonly used, let's now compare them in terms of: *heat vs. light, renewable energy* , and *true bulb cost.*

Heat vs. Light

We buy light bulbs for lighting, not heating. To help you choose the most efficient bulb for energy released as light and not as heat, let's take a look at the illustration blow.

As we discussed, the incandescent light bulb produces 90% of its energy as heat instead of light. That is very inefficient, to say the least. It only has to be on for a short time and it is too hot to touch without being burned.

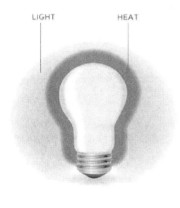

INCANDESCENT BULBS RELEASE 90% OF THEIR ENERGY AS HEAT AND ARE HOT TO THE TOUCH.

A CFL RELEASES ABOUT 80% OF ITS ENERGY AS HEAT. IT CAN BE VERY WARM TO THE TOUCH.

LED LIGHTING RELEASES AS LITTLE AS 46% OF ITS ENERGY AS HEAT BACKWARDS INTO A HEAT SINK AND IS COOL TO THE TOUCH.

Figure 1.19 Heat & Light emission comparison of CFL, LED and Incandescent

CFL radiates 80% of its energy as heat and can be very warm to the touch. It has been marketed as a "green" energy-efficient bulb because it uses 77% less energy than a comparable incandescent bulb for the same amount of lumens (light output). It is warm to the touch because it releases 4/5th of its energy as heat.

The best lighting efficiency, at this time, is LED. It releases 47% to 64% of its energy as light. Even if the LED you buy is at the low end of that range, it is a significant increase in efficiency over incandescent lighting and CFL. It is cool to the touch, mainly because of this efficiency and the fact that it also has a heat sink.

In terms of heat vs. light, the basic rule of thumb in lighting is this: if it is cool to the touch after being used for 5 minutes, you can be sure that it is a reasonably efficient bulb.

Renewable Energy (Illuminating Example)

How many light bulbs do you have in your home? 30 is a pretty average number. Again, let's compare three types of lighting: the standard incandescent bulb, the compact fluorescent bulb (CFL), and the LED bulb. In this example, all three bulbs put out approximately the same amount of light, 850 lumens.

30 x 60 watt Incandescent bulbs = 1,800 watts used to light your home

Or

30 x 14 watt CFL bulbs = 420 watts used to light your home

Or

30 x 10 watt LED bulbs = 300 watts used to light your home

So...

The material cost of a solar array system runs about $5/watt (The installation could increase that to about $8/watt). So, here is some simple math….

The Incandescent bulbs usage of 1,800 watts x $5/watt = $9,000 array cost

Or

The CFL bulbs usage of 420 watts x $5/watt = $2,100 array cost

Or

The LED bulbs usage of 300 watts x $5/watt = $1500 array cost

How much do you want to spend on your renewable system? You can change the cost dramatically by just changing your light bulbs.

True Bulb Cost

The true cost of lighting choice is clearly illustrated in the following chart. It compares the cost of 50,000 hours use of incandescent, CFL and LED bulbs of equivalent lumens. This includes bulb lifespan, bulb cost, wattage used, electricity cost, and the cost for 30 bulbs (the average amount of bulbs in a home).

BULB COMPARISON Figure 1.20	LED	CFL	INCANDESCENT
Lifespan - How long will the bulb last?	50,000 hours	10,000 hours	1,200 hours
Watts per bulb - Wattage equiv. at 60w	10	14	60
Cost per bulb	$29.98	$2.98	$1.25
KWh of electricity used of 50,000 hours	500	700	3,000
Electricity cost - @ $0.10 per KWh	$50.00	$70.00	$300.00
Bulbs needed for 50,000 hours of usage	1	5	42
Equivalent 50,000 hour bulb expense	$29.98	$14.90	$52.50
Cost of 50,000 hrs of electricity + bulb(s)	$79.98	$84.90	$352.50
Cost for 30 household bulbs	$899.40	$447.00	$1,575.00
Cost for 50,000 hours of electricity for 30 household bulbs	$1,500.00	$2,100.00	$9,000.00
Cost of 50,000 hrs of electricity +bulb(s) use for 30 household bulbs	$2,399.40	$2,547.00	$10,575.00
Savings by switching from incandescent	$8,175.60	$8,028.00	0

Green Note: The European Union, as of September 1, 2009, has banned the sale of any light bulb that uses over 75 watts. The U.S. is slated to ban incandescent bulbs in 2012.

BULB COMPARISON - continued	LED	CFL	INCANDESCENT
Ecology and Environment	Very Friendly/ min. issues	Damaging Mercury/ Argon	Damaging
Heat Issue	Least	Ballast Heat	Largest (90%)
Light Control	Most Control	Least Control	Variable Control
Maintenance	Zero	Ballast Issue	Replacement
Weather Temperature Changes	Not Sensitive	Sensitive	Some Sensitive
Effective Lumens	Equivalent	Equivalent	Equivalent

In summary, we have discussed the most commonly used light bulbs in homes today. By simply exchanging the high wattage bulbs for more efficient ones, you can make a big difference in the amount of electricity that you use in your home. This translates to a smaller renewable energy system, conserving electricity, avoiding constant bulb maintenance, and still enjoying a bright home.

Tubular Daylighting Devices

The new Tubular Daylighting Devices (TDD) are not the skylights that your parents or grandparents had in their homes, although, they serve the same purpose. TDDs put out incredible light, providing as much light as you would expect from a skylight many times its size. They are easy to install, reduce electricity usage, and can put a smile on your face in the winter.

Installation is fast, clean, and easy. They require no structural reframing, furring, dry walling, or painting. A professional can install

Figure 1.21 Cut-away of a Tubular Daylight

the product in less than two hours and most Do-It-Yourselfers can finish the project in less than a day. They can go almost anywhere. The compact and flexible design of TDDs allow them to be installed in just about any room, including rooms without direct roof access and smaller spaces where day-lighting would usually not be an option. They don't leak, nor do they allow heat to escape in the winter as their predecessors do.

Once you have paid the price of materials and installation, they provide you unending day-time lighting. This allows you to switch off electric lights during the day, which provides savings on energy bills.

It has been shown in several prominent studies that increasing daily exposure to natural light can enhance mental and physical well-being, boost concentration and energy levels, and provide a variety of other unexpected benefits.

Whether you are remodeling or building a new home, natural lighting is a cost-effective form of lighting that will help you to reduce your energy consumption and provide extra light on short winter days.

Important Points:

- **Choose efficient light bulbs that compare to incandescent light output (lumens) but use less electricity (watts).**

- **The lighting choice will save you money on electricity and maintenance over time.**

- **Efficient lighting will lower the const of renewable energy systems.**

APPLIANCES
Are All Appliances Created Equal?

APPLIANCES

Your Questions:

- **Does it really pay to purchase more expensive Energy Star appliances?**
- **How much of my home's energy costs are from my appliances?**
- **Will Energy Star appliances help to save money on the purchase of renewable energy systems?**

Before you purchase an appliance, remember that it has two price tags: what you pay to take it home and what you pay for the electricity it uses. Energy Star appliances have advanced technologies that can use 10% to 50% less electricity than standard models.

Figure 1.22 Get rid of your old, inefficient refrigerator for an efficient Energy Star refrigerator.

For example, an Energy Star qualified refrigerator can draw as little as 100 watts of electricity, while a standard refrigerator can draw as much as 500 watts. The money you save on your utility bills can make up for the difference in the purchase price. To find out how much you can save, go to: http://www.energystar.gov/index.cfm?fuseaction=refrig.calculator

Purchasing an Energy Star qualified model rather than a non-qualified model will save you an average of $50 a year on your utility bills. Over the life of your new washer, you'll save enough money to pay for the matching dryer. Washers built before 1998 are significantly less efficient than newer models. If your washer is more than 10 years old, you're paying about $135 more each year on your utility bill than you would if you owned a new Energy Star qualified model.

"All Energy Star home appliances must meet the Appliance Standards Program set by the US Department of Energy (DOE). Test results are printed on the yellow Energy Guide label, which manufacturers are required to display on many appliances. This label estimates how much energy the appliance uses, compares energy use of similar products, and lists approximate annual operating costs. Your exact costs will depend on local utility rates and the type and source of your energy." (source DOE)

In the U.S., appliances in a typical home are responsible for about 20% of your energy bills. These appliances are dishwashers, clothes washers, dryers, TVs, refrigerators, water heaters, and more. By shopping for appliances with the Energy Star label and turning off your appliances when they're not in use, you can achieve a real savings on your monthly bills and in purchasing your renewable energy systems.

Figure 1.23 Energy Star clothes washer and dryer set.

Many appliances continue to draw a small amount of power when they are switched off. These "phantom" loads occur in most appliances that use electricity, such as VCRs, televisions, stereos, computers, and kitchen appliances. In the average home, 75% of the electricity used to power home electronics is consumed while the products are turned off. This can be avoided by unplugging the appliance or using a power strip and using the switch on the power strip to cut all power to the appliance.

Example of Savings;

The average NON Energy Star Refrigerator uses about 500 watts

Or

Some Energy Star Refrigerators use about 100 watts.

So…

The material cost of a solar array system runs about $5/watt (The installation could increase that to about $8/watt). So, this is simple math….

$$500 \text{ watts x } \$5/\text{watt} = \$2{,}500 \text{ array cost}$$
Or
$$100 \text{ watts x } \$5/\text{watt} = \$500 \text{ array cost}$$

It is very cool (pun intended) how much you can save on a renewable energy system by choosing an Energy Star refrigerator. Appliances matter when you have the goal of having an energy-efficient home that is also powered by your own renewable energy system. Energy Star appliances last longer and lower electricity usage. This can mean a smaller price tag for a solar array, wind generator, or back-up generator.

> **Important Points:**
>
> - **Energy Star appliances are better built and longer lasting.**
> - **Energy Star appliances pay for themselves by lowering energy consumption.**
> - **Energy Star appliances use less energy and save you money when purchasing renewable energy systems.**

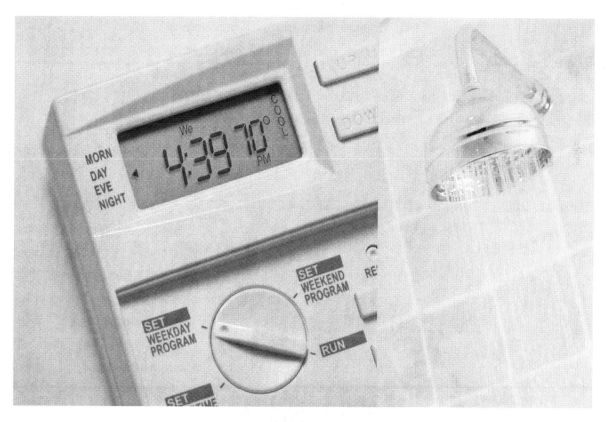

YOUR HOME'S SYSTEMS
Heating & Cooling and Hot Water Production

YOUR HOME'S HEATING & COOLING AND HOT WATER SYSTEMS

Figure 1.24 HVAC Thermostat

With advancing technologies available today, no one should be wasting money on antiquated and inefficient Heating, Ventilation & Air Conditioning (HVAC) or hot water heating equipment. In this chapter, we introduce you to some of the best HVAC and water heating systems available for retro fit or new construction. We will explain these systems starting with the best first, but they are all very efficient and have Energy Star ratings and tax benefits.

Figure 1.25 Hot water

SPACE HEATING & COOLING (HVAC)

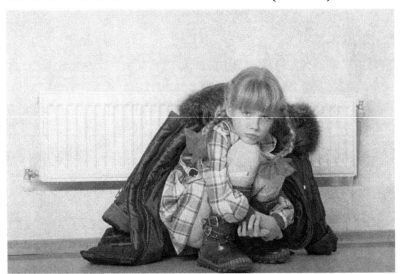

Figure 1.26 Do you have a home that is comfortable to live in?

Having a comfortable air temperature in our homes at an affordable price is crucial to conserving both money and energy. Your home should be a place of comfort and rest in an ever-demanding world. We could discuss many types of heating and cooling systems, but we wanted to share two of the best system types available. Our top choice is a Ground Source Heating and Cooling system for its extremely efficient transfer of energy, sometimes approaching 500% efficiency. The second type of system we suggest, if you cannot have the first, is a hybrid heating and cooling system, efficiently using electricity *and* natural gas or propane.

GROUND SOURCE HEATING AND COOLING

Your Questions:

- **Is ground source heating and cooling the same as geothermal?**
- **How can I use the earth to heat and cool my home?**
- **How does a ground source heat pump transfer energy from the ground or water to heat and cool my home?**
- **What are the components of these systems?**

Ground source heating and cooling is also known as geothermal heating and cooling. These terms are alternately used in this section. Both names let you know that you are using the ground to heat and cool your home. This type of HVAC has nothing to do with geothermal hot water like you see at Yellow Stone Park. Ground source heating and cooling is simply drawing on the earth's constant temperature to heat and cool your home. In this section, we will explain in-depth how geothermal heating and cooling works by examining what a heat exchanger is and different types of ground source heat pumps.

GROUND SOURCE HEAT EXCHANGE LOOP

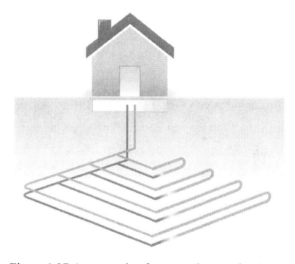

Figure 1.27 An example of a ground source heat exchange loop

A geothermal heating and cooling system works differently than other systems. Unlike conventional systems, ground source heating and cooling systems do not use a fuel, electric heating elements, or the outside air to produce heat, but use the temperature of the ground. At a depth of 4 to 8 feet, the earth has a nearly constant temperature of 45°F to 75°F, depending on where you live. The heat is exchanged or transferred to and from water or an antifreeze solution. As you will see in the next section, the water can be circulated through polyethylene pipes in the ground or body of water where the heat exchange happens, or can be pulled from a well. This process is reversed in the summer for air conditioning by extracting heat from the home and circulating it to the ground.

It is the most efficient and environmentally friendly home heating and cooling method available. For every one unit of electricity, you get 3 to 5 units of heat or 300% to 500% efficiency. With a gas furnace, for every one unit of fuel, you get .8 to .96 units of heat or 80% to 96% efficiency. As an added benefit, a ground source system can provide up to 70% of your hot water free by sending excess heat to your water heater.

If you are new to the topic of ground source heating and cooling, we encourage you to go to our website, www.MyPowerfulHome.com, to watch an introductory video along with reading a great article on this topic as well. You can "Ask Keith" questions and he will email you the answer or make a video answer to put on the site. Also, search the U.S. government site Geothermal Heat Pumps at: www.energysavers.gov.

Important note: The federal tax credit for geothermal heat pumps or a complete heating and cooling system and installation are the same as that for any renewable energy system. This credit reduces the total cost by 30%.

Ground Heat Exchanger - part of the system.

To help you understand this underground heating/cooling exchange, let's compare it with a conventional system you may be familiar with, an air source heating and cooling system. Air source heating and cooling uses a furnace air system inside your home along with an air source heat pump outside. That noisy unit outside your kitchen window or in the back yard

Figure 1.28 Air source heat pump.

has to extract heat from your home to cool it and move this heat outside.

When heating your home, that same noisy heat pump outside has to extract heat out of the air when it is very cold outside and pump it into your furnace/air handler for distribution. When it gets very cold outside, your heat pump becomes too inefficient in extracting heat, so your electric heat strips or your fuel furnace takes over heating your home and you spend more money.

As the outside temperatures cycle seasonally, remember that the ground temperature stays nearly constant 45°F to 75°F. So, what do you think is more efficient? Moving heat in or out of ground temperatures that range from 45°F to 75°F, or in or out of outside air temperatures of 0°F to +100°F? Would you rather pay much more for fuel or electricity to heat and cool your home or take advantage of the earth's constant ground temperature?

Different Types of Ground Source Heat Exchangers
The two ways to move heat energy between the ground and your home are through a *closed loop* or *open loop* system.

A Closed Loop system may consist of several pipe loops, coiled like a slinky, or it may simply be vertical or horizontal.

- Slinky and horizontal loops are buried from 4 feet to 8 feet deep. The length of pipe can vary depending on the moisture in the soil and the soil type. These factors are important because they affect the transfer of heat between the pipe and the earth. Special polyethylene pipe must be used, because it has the correct properties to allow the transfer of energy. In cold climates, a propylene glycol or methanol antifreeze mixed with water is used to avoid freezing. Pumps force this mixture through the pipe creating turbulence in the fluid for an efficient transfer of energy to and from the earth.

- Vertical drilling can be used when you have very little land available and can be installed on most properties. Of the closed loop options, vertical loops are usually the most expensive; however, new drilling technologies are developing that can make this loop cheaper and easier to install.

- Ponds provide a great energy exchange at depths of around 8 feet or more. The best transfer rate is possible when your loop is in direct contact with water and fewer pipes are needed. There are prefabricated stainless steel devices that transfer heat using less surface area. These are placed at the bottom of a pond or body of water. One of these is sold by the brand name, Slim Jim Geo Lake Plate.

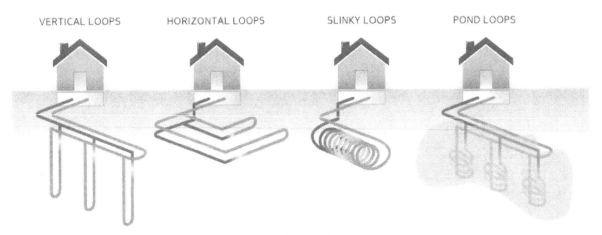

VERTICAL LOOPS   HORIZONTAL LOOPS   SLINKY LOOPS   POND LOOPS

Figure 1.29 Four types of closed loop ground source heat exchangers.

When designing your ground loop system, knowing the available land and soil type is very important for a properly working system. The more moisture in the soil, the better heat is transferred. Soil types exchange heat and cold at different rates or with different efficiency, so make sure to have your system professionally designed.

An Open Loop or pump and dump system uses well water that also has a constant temperature and circulates this water through your unit in your home to accomplish the heat energy transfer. The open loop generally needs to dump or discharge 2-9 gallons of water per minute when the unit is heating or cooling. Some people discharge this water into a stream, reinjection well, pond,

Figure 1.30 Open loop

canal, storage holding tank for irrigation, etc. Open loop systems can be the least expensive of all ground source systems to install.

Types of Ground Source Heat Pumps (GSHP)

A ground source heat pump, that looks like a furnace, combines two specific functions of heat exchange. First, fluid is pumped through pipes buried in the earth or from a well to a GSHP unit inside your home. Through this pumping/circulating process, heat is moved from the earth to the home or from the home to the earth through this unit. Secondly, inside the GSHP, this heat is collected and compressed to a high enough temperature to heat your home. This compressing process is reversed for cooling and heat is moved from your home, through your GSHP unit, and back to the earth. These two processes give us a mechanical unit called a "ground source heat pump." First, we will examine four GSHPs: *water to air, water-to-water, combination units,* and *split system units.* Then, we will present important *options to consider when installing a GSHP.*

Water to Air GSHP

A water to air ground source heat pump transfers the heat to and from the earth by circulating an antifreeze/water solution through a copper coil inside the GSHP heat exchanger. This copper coil is also wound tightly, touching another copper coil that is full of high pressure Freon.

The Freon coil extracts this heat to and from the ground loop and uses a compressor to move this energy for conditioning your home. The compressed heat/cold is pumped through coils inside your GSHP that radiates this energy to be carried by fan-forced air to the home.

To sum it up, there are three transfers taking place:

First, the open or closed ground loop circulates fluid through the ground to transfer heat/cold energy. The sun radiates this energy to the ground, and it is from the ground that it is utilized.

Secondly, inside the furnace, two copper coils transfer this heat energy through a heat exchange process to a compressor which significantly raises or lowers the temperature of this fluid.

Thirdly, this heat energy is then pumped from the compressor through coils located in the path of an air handler to circulate this heat energy in to or out of your home.

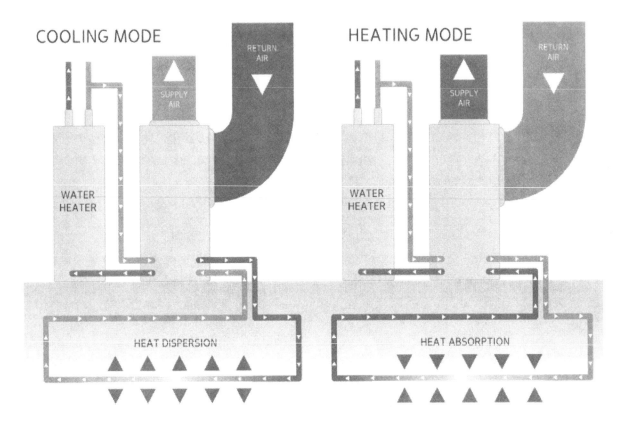

Figure 1.31 Water to Air Ground source heat pump exchange process. It shows that free excess heat, from the desuperheater, can be used to heat nearly all of you domestic water.

While all of this heat energy is being transferred, excess heat from the compressor is being produced. This excess heat can be circulated through your hot water heater with a desuperheater (internal hot water heater), giving you free hot water. This hot water is free because you need to heat and cool your home regardless of your hot water needs. The exchange is so efficient that it requires far less energy to transfer the heat than conventional systems. At the end of this chapter, we provide some more information on desuperheaters, so read on!

Water-to-water GSHP

A water-to-water heat pump uses the same transfer process we previously mentioned, with one exception. Instead of the third transfer from radiator-like coils to the air, heated or chilled water is pumped to air handler fan units in different areas or levels of the home. This can save money and add to your system's efficiency by reducing the amount of ductwork and heating/cooling loss by pushing air long distances. It is more efficient to pump water than it is to push air long distances or through multiple levels in the home. See figure 1.34.

The hot or cold water, from the water-to-water unit, can pass into a buffer tank before being pumped to remote air handler fan units as mentioned above. A buffer tank is an insulated reservoir and allows you to store a lot of hot or cold water. The hot water can then be used to indirectly preheat domestic water before it goes into the hot water heater. This can be a tank type or an instantaneous water heater, making it far more efficient for heavy hot water use.

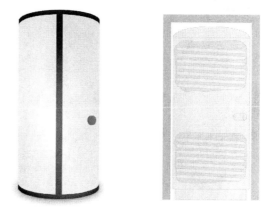

Figure 1.32 Buffer tank with internal view.

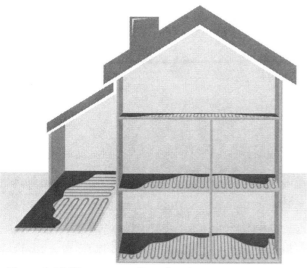

Figure 1.33 Hot water radiant flooring.

A buffer tank can be used to preheat water for pool/spa heating or for radiant floor heating. Of course, your main heat will come from the GSHP, but using radiant floor heating to warm your floors is a great answer to those cold wood or tile floors. It is a wonderfully comfortable heat if it is properly controlled. You are actually taking the heat out of the ground and putting it under your feet in your home.

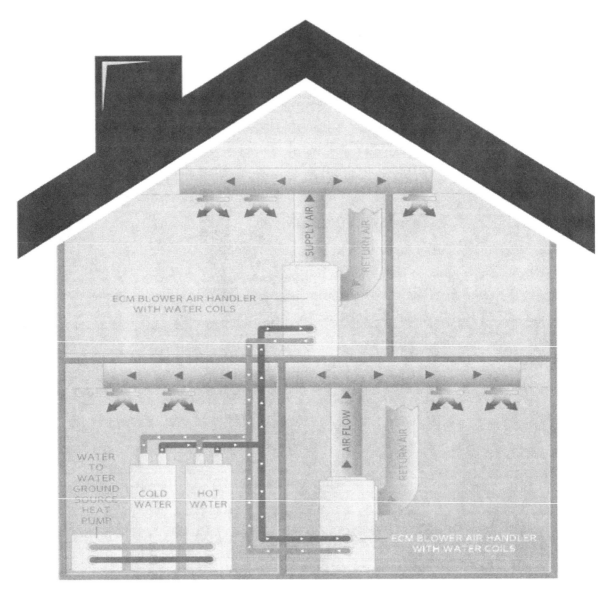

Figure 1.34 An Example of a complete water-to-water GSHP home system.

Combination Water-to-Water and Water to Air GSHP

Combination units can give you everything previously mentioned all in one unit. Combination units look like your furnace in the basement, but they are multi-task units. These units can be set to prioritize according to a desired demand preference. For example, if you are air conditioning and you need a lot of extra hot water, the unit can briefly shift its function to meet the demand. Combination units are becoming popular for people who enjoy the comforts of radiant floor heating and also want forced air for air conditioning and air quality. When you are in those in-between seasons, you may need heating and cooling in the same day, and a combination unit can be a great choice.

For the technically inclined, here is an internal picture of a GT Combo Unit from Geocomfort with component explanation below.

Figure 1.35 Geocomfort combination water-to-water GSHP.

1. Coated Air Coils For Extended Life
2. Corrosion-Proof Stainless Steel Drain Pans
3. Copleland Ultra Tech Two-Stage Scroll Compressor
4. Digital Controls
5. Optional Hot Water Generator (Desuperheater)
6. ECM Blower Motors
7. Optional Auxiliary Electric Heaters
8. Condensate Overflow Sensor
9. Hydronic Heating (Built-In Water-to-Water Section)

Figure 1.36 Geocomfort combination water-to-water GSHP internal.

Split System GSHP

Split systems can be great for some retrofit installations, especially in areas where it doesn't freeze and you can put your GSHP unit outside. This can be done by replacing your outside air source heat pump with a ground source heat pump and ground loop pump pack. You would then remove the existing coil in your furnace/air handler and replace it with a coil matching your new ground source heat pump. The existing electrical power and control wiring may all be used (if they are the right size) with your new system. Next, install your ground loop heat exchanger, open or closed loop, as discussed earlier, and you have your renewable energy system.

Options with Ground Source Heat Pumps

There are three options, we strongly suggest, to include with your GSHP. These are: An ECM blower, a two-stage compressor, and a desuperheater.

- ECM (electric commutated motor)

  For heating and cooling purposes, an ECM is essentially a variable speed fan motor. It is located in the air handler and runs the fan that pushes the conditioned air into your home. This option gives you a motor that will vary its speed according to the desired amount of heating or cooling. These motors start slowly, then increase as the airflow system demands. This increases the life of the motor, uses only the electricity it needs, and saves on money and maintenance. A computer system controls the ECM to increase or decrease the airflow to accomplish the task. You can set your system fan to slowly sweep and condition the air when there is no heating or cooling demand.

  Note: While this air is being constantly, almost undetectably, moved, ultraviolet lights (if installed) will kill germs and bacteria. Also, with an electronic air cleaner or a high quality air filter installed, you can get rid of circulating dust particles. This is another part of your system that will help those who have allergy problems or compromised immunity.

- Two-stage Compressors

  Your system should be designed to meet extreme weather demands. In many areas, temperatures vary widely throughout the year, yet have six months of mild temperatures. Mild seasons require less heating/cooling capacity than extreme seasons. To meet the varying heating and cooling demands with temperature change, be sure that your system has a two-stage compressor.

Two-stage compressors will use the first stage, or about ½ the capability of your system, to satisfy the demand during these moderate times. This also gives your ground loop system and your GSHP heat exchanger greater efficiency. To say the least, your whole system will be very efficient during the first stage. It only makes sense to have a system that uses about half of the energy to run when the climate is mild.

When the temperature differential becomes too great, the system will detect this and ramp up to full capacity, bringing the second stage to work until the demand is satisfied. While all this is happening your ECM fan blower is also matching the airflow needs of the system.

- Desuperheater – FREE hot water!
  This option uses the excess super heat from the compressor during both heating and cooling cycles to meet hot water needs. A desuperheater heat exchanger (coil) uses this excess heat to heat domestic water, while the system is still satisfying its heating and cooling needs. Remember, the unit has to run anyway to heat and cool the home, so this water heating is practically free. Some manufactures claim that these systems will provide all of your hot water needs. We still recommend additional hot water heating, like solar, electric or instantaneous gas water heaters, for households with heavy hot water usage. Go to energysavers.gov search *heat pump features.*

In this introductory and yet detailed section on ground source heating and cooling, we have described: how geothermal or ground source heating and cooling works, the different types of ground heat exchangers, and the types of heat pumps with option recommendations. The amount of detail might have been too much for some, but adequate for others. Either way, we have put it in terms that we feel can be understood by the average homeowner.

Ground source heating and cooling is the most efficient way to heat and cool your home. With 300% to 500% efficiency and the benefit of free hot water, your home energy requirements can be significantly reduced. This means that a smaller renewable electrical system will be needed to cover the energy required for heating and cooling and hot water production.

Important Points:

- **Ground source heating and cooling (geothermal heating and cooling) utilizes the nearly constant, temperature of the earth (45°F to 75°F) at a depth of 6' to 8' and greater.**

- **Ground source heating and cooling is 300% to 500% efficient.**

- **Free hot water is a side benefit when you have a desuperheater, which reduces home energy requirements.**

- **Make sure to include an ECM blower, two-stage compressor, and a desuperheater with your ground source heat pump system.**

HYBRID HEATING & COOLING

> ### *Your Questions:*
>
> - **What is hybrid heating and cooling?**
> - **What is the difference between a hybrid system and a ground source system?**
> - **Are their tax incentives available for hybrid systems?**

Hybrid heating and cooling systems commonly use both electricity and natural gas or sometimes, they incorporate propane instead of natural gas. These systems operate very efficiently because they use an efficient electric air source heat pump combined with an efficient natural gas/propane furnace working together to heat and cool your home.

An efficient hybrid system should include a multistage air source heat pump, or a heat pump with at least two stages. This allows the heat pump to extract heat to and from the home in different stages. The first stage will satisfy the heating/cooling demand inside your home during mild outside temperatures. When the outside temperatures become extreme, the compressor's second stage will activate, using more electricity to satisfy the demand. During the cooling season, these two stages will satisfy the cooling need. Your heat pump should have an efficiency rating of at least 14 SEER, (seasonal energy efficiency ratio)which is the unit's efficiency under a range of weather conditions. The higher the SEER rating, the more you save on electricity costs.

Heating inside space with an outside air-source heat pump is a different story. Most heat pumps lose their efficiency in extracting heat out of the air when the outside air gets below 32°F. When it is this cold outside, an additional heat source is needed. This is where the term "hybrid" comes into play. At 32°F, your natural gas/propane furnace is activated and begins supplying heat. These furnaces should be very efficient and have an Annual Fuel Utilization Efficiency (AFUE) of 95 or greater (This is 95% efficiency). They should be efficient enough to have PVC for venting instead of a metal flue pipe. A gas furnace in a hybrid system should be modulating. This means that the size of the flame in your furnace should only increase as the demand for heating increases. Most hybrid systems are computer controlled, measuring inside and outside temperatures and then adjusting the system accordingly for the greatest efficiency.

The difference between a ground source system and a hybrid system is very simple. First, a ground source system requires less maintenance and has fewer moving parts. Secondly, a ground source system does not require an outside heat pump and an inside furnace with a fuel source. One inside unit contains all that is required to heat and cool your home. Thirdly, and most importantly, it is more efficient to transfer heat energy from the constant temperature of the earth than from the varying outside air temperatures.

While a hybrid system does not have all the advantages of a geothermal unit, it is the second most efficient option. Many utility companies offer incentives for efficient hybrid systems. You can also enjoy some tax benefits from state and federal governments in the U.S. for qualified hybrid heating and cooling systems. The Energy Star website does not clearly indicate tax benefits for a hybrid system, so contact a heating and cooling contractor in your area that can walk you through the requirements and give you documents for tax purposes.

Important Points:

- **Be sure your outside air source heat pump in your hybrid system has at least two stages.**

- **Your outside heat pump should have at least a 14 SEER rating.**

- **Your furnace should have a modulating flame increasing or decreasing in heat output as required for the greatest efficiency of the entire system.**

- **Make sure that your furnace has an AFUE of 95 (95% efficiency) and is vented with PVC plastic pipe.**

HOT WATER PRODUCTION

Your Questions:

- **How much of my water will a desuperheater heat?**
- **How do I integrate a solar water heater with my existing system?**
- **How much water do I need to store for my family?**
- **What types of solar water heaters are most common?**
- **How does a Heat Pump Water Heater (HPWH) work?**
- **Will a HPWH replace my existing water heater?**
- **What are some applications for instantaneous water heaters?**
- **Will an instantaneous water heater supply all of my hot water needs?**
- **What are the best features to get in an instantaneous water heater?**

There are several different ways to heat water for your home. In this chapter, we will encourage you to think outside of the tank. We will discuss four alternative options for complete or assisted water heating: *Desuperheaters, Solar Hot Water, Heat Pump Water Heaters (HPWH) and Instantaneous Water Heaters.*

Figure 1.37 Enjoy several way to produce hot water in your home.

DESUPERHEATERS

You can also refer to the ground source heat pump section of this book to learn about desuperheaters. This technology uses excess heat from your geothermal heating and cooling unit to provide up to 70% of your hot water free. When operating, a GSHP desuperheater raises the temperature of the water inside your existing water heater or in a buffer tank before it enters your water heater. This occurs by raising the temperature of the water each time it circulates through the ground source heat pump in the process of heating and cooling your home. You should never avoid this option with a ground source heating/cooling system because it is simply free hot water from excess heat. To add a desuperheater component to your GSHP costs about $300; however, some companies include it in the basic cost of a GSHP.

SOLAR WATER HEATING

Figure 1.38 Solar thermal panel for water heating.

We all like hot showers, hot dishwater, hot water to wash our clothes, and even a hot drink. Solar energy can provide most or even all of that hot water for your home. Solar water heaters have been in use in America since the 1920s. Even with a long history of use, most people do not have a solar water heater or know exactly how they work. We will discuss how they work, diagram an active indirect system, and expand on the topic of hot water storage.

How Solar Water Heating Works

A solar water heater uses the sun's energy to preheat household water generally before it enters the conventional gas (or electric) water heater. Standard residential solar water heating systems reduce the need for water heating by about two thirds. These systems combined with a desuperheater in a ground source heat pump, instead of a conventional water heater, can reduce your water heating costs to almost nothing. At the very least, your electric or gas bill will be significantly reduced and you won't have to count the minutes when someone takes a long shower.

Most of the solar water heating systems have two main parts, a solar collector and a storage tank. The most common collector used in home solar water heating systems is a *flat plate collector*. A flat plate collector is a shallow rectangular box that has a clear glazed surface.

Under the glazing is an absorber plate that has many small copper tubes running through it that collect into a larger copper manifold. This type of collector is generally 4' wide by 8' long and 4" to 6" deep, but can also come in other similar sizes.

The sun's energy is absorbed by the absorber plate, which heats up. The heat of the plate transfers to the copper tubing that warms circulating water. To keep the water from freezing in colder climates, a food grade antifreeze can be added in a separate loop. This heated water/antifreeze is then held in the storage tank ready for use. It is a simple but effective technology with an efficiency rating of about 54%.

FLAT-PLATE CROSS-CUT

Figure 1.39 Flat plate solar collector.

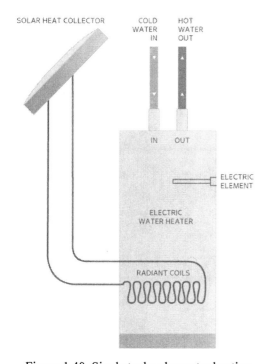

Figure 1.40 Single tank solar water heating

Hot Water Storage

The old adage, "Make hay while the sun shines" also applies to all things in solar energy. In order to "make hay" or make use of the sun while it shines, you must have a way to store the energy or hot water it creates for use when the sun is gone. A small (50 to 60 gallon) storage tank should be sufficient for one to three people. A medium (80 gallon) storage tank works well for three to four people. A larger (120 gallon) tank suits four to six people. You may choose to have a *single tank system* or a *two-tank system* to have enough storage for your household demand.

Single Tank System

A single tank system works well when there is not a high demand for hot water. Typically, an elec-

49

tric water heater with solar heating coils is used, and solar heated water is stored in this water heater tank. When you have a sunny day, water is continuously circulated through a solar panel to your water heater until a desired temperature is reached. Electrical heating elements maintain a desired temperature in this tank at night and can assist in water heating during periods of high hot water demand. Set the temperature on the tank thermostat for electrical water heating 10 to 15 degrees below the temperature setting for solar water heating. This will let the sun's free energy be your primary source of water heating and still give you an extra boost when needed. See figure 1.40.

Two-tank System

Let's look at a two-tank system for top notch efficiency. A separate tank is used to preheat the water from the solar storage tank before it enters the conventional gas or electric water heater. For example, if your solar water heater heated up an 80 gallon tank to 120°F during the day, and the next morning, the tank had cooled to 95°F, then the secondary water heater tank would only have to heat it up 10°-15° for use. A two-tank system is usually considered the best option because of the storage capacity of two tanks and greater heating efficiency.

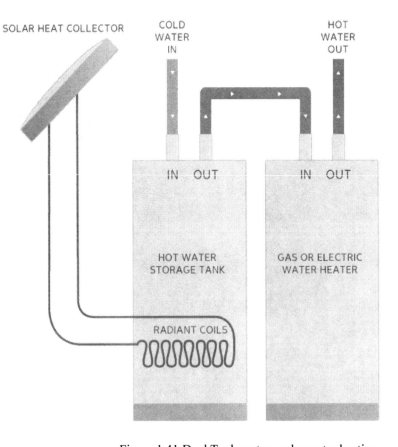

Figure 1.41 Dual Tank system solar water heating.

50

Active Solar Water Heaters

Solar water heating systems can be either passive or active. A passive system does not use pumps to circulate the heated fluid. It uses only the principle of rising heat to circulate fluid passively within the system. This type of system is less efficient and less commonly used than are active systems. Because of this, we will not go in to passive solar water heaters in this book.

Active solar water heaters rely on electric pumps and controllers to circulate water or anti-freeze solution through the collectors.

The following are the two types of active solar water-heating systems: *Active Direct* and *Active Indirect*.

1. Direct-circulation systems use pumps to circulate pressurized potable (drinkable) water directly through the collectors. These systems are used in areas that do not freeze for long periods and do not have hard or acidic water that can degrade or clog the loop over time.

2. Indirect-circulation systems pump an antifreeze solution through a collector, and then, to a storage tank. In the tank, heat exchanger coils transfer the heat from the solution to the potable water. Your drinking water and the solar water never mix because they are in different loops. (See the following page for an Active Indirect Solar Water-heating diagram.)

Solar heating has a payback period that can vary from four to ten years. It may be more useful to think of solar water heating as an investment that yields an annual return, much as a bank savings account provides interest. A solar water heater will generate savings that can equal a bank account generating a twenty percent (20%) annual return, and the savings are not taxed as income, as is the interest you earn at the bank.

In terms of solar energy applications, solar water heating is the most inexpensive and effect-tive. Active systems can cost up to $6,000 or more depending on size. Federal and some state tax incentives and rebates can drop that price by 30% to 50% or more. Solar water heater federal tax credits are higher than for any other type of water heater because it is actually placed in the Solar System tax benefits category.

Figure 1.42 shows an example of an Active Indirect One-Tank system. This shows an electrical element as a backup heat source, but this could be gas or a desuperheater as well. The controller shown in the diagram shuts off the system at night so cold solar panels do not cool the storage tank when there is no sun energy. In the morning, the controller turns on the pump, only when the solar panel reaches the desired temperature.

ACTIVE INDIRECT ONE-TANK SYSTEM

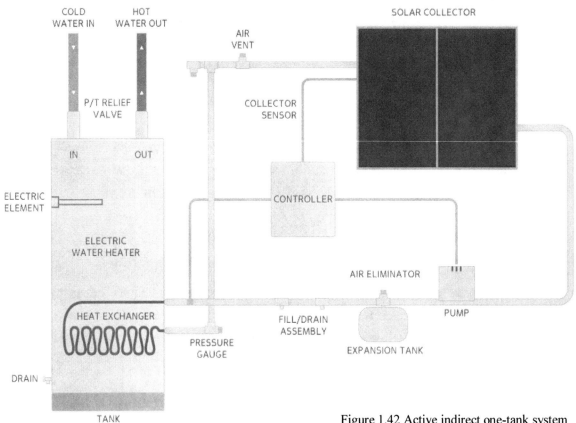

Figure 1.42 Active indirect one-tank system

Green Note: Solar water heating is a technology that every homeowner can use to save on utility bills. In a year, the average household of four will consume enough energy to heat water to fuel a midsized car for 20,000 miles.

HEAT PUMP WATER HEATERS (HPWH)

The DOE is also promoting a new indoor air source heat pump water heater. It could be a good choice if you don't see a ground source heating and cooling unit with a desuperheater in your future. It can also work well in conjunction with a solar water heater, especially if you do not have gas available.

The new HPWH (heat pump water heater) extracts heat out of the air inside your home. After extracting this heat, it is then compressed and put through a coiled heat exchanger in the tank. This compressed energy from the air efficiently transfers the heat, generating hot water for your home. It's kind of like your freezer, except the process is reversed, rejecting cold into the air instead of heat. Put a HPWH and a freezer side by side and this will help equalize the rejected hot/cold air. The HPWH has a 200% efficiency rating and could replace an electric water heater. Energy Savers: Heat Pump Water Heaters. For federal tax credits for this type of water heater got to www.EnergyStar.org.

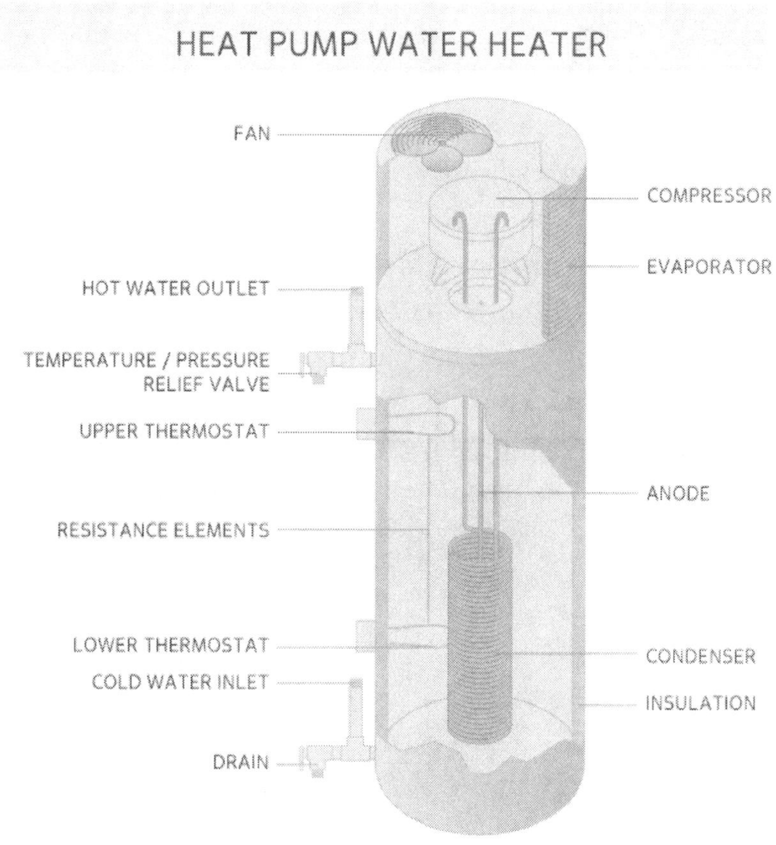

HEAT PUMP WATER HEATER

FAN
COMPRESSOR
EVAPORATOR
HOT WATER OUTLET
TEMPERATURE / PRESSURE RELIEF VALVE
UPPER THERMOSTAT
ANODE
RESISTANCE ELEMENTS
LOWER THERMOSTAT
COLD WATER INLET
CONDENSER
INSULATION
DRAIN

Figure 1.43 Heat Pump Water Heater and components

INSTANTANEOUS WATER HEATERS

Instantaneous or demand water heaters are tankless and can be gas or electric. These have many applications other than being used under sinks to avoid a long wait for hot water to arrive while you are washing your hands. They can be used to increase the temperature of stored hot water efficiently. They can also be used as backup heat, to assist a ground source heat pump system, or to raise the water temperature from a solar hot water storage tank. (See the line diagram at the end of this section.)

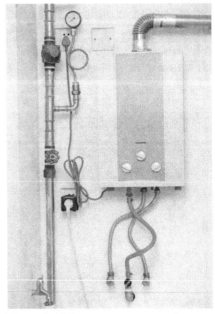

Figure 1.44 Instantaneous water heater

Sometimes, these water heaters are expected to supply all of the water heating needs. People end up disappointed because instantaneous water heating, by itself, usually does not supply all the hot water for a family. It is difficult to heat all of your water instantaneously when everyone is showering, clothes are being washed, and the dishwasher is in use.

We are seeing some super efficient instantaneous water heaters coming to market. The most efficient instantaneous water heaters are vented with PVC, not metal, and vary in heating output or modulate with the water heating demand. To wash your hands, you don't need to have 200,000 btu when you only need a quick shot of 30,000 btu. Go to ahridirectory.org and click in the upper right corner to find qualified water heating products that meet the tax credit efficiency minimum.

Important Points:

Desuperheater

- **Hook up your desuperheater to your existing water heater.**
- **You should always choose this option with a ground source heat pump.**

Solar Water Heater

- **Size your solar hot water storage tank properly so you utilize as much of the sun's free energy as possible.**
- **For the best results, hook up your electric or gas water heater after your solar hot water storage tank.**
- **Do not circulate water through your solar water heater panel when it is not generating hot water from the sun.**

Heat Pump Water Heater

- **A HPWH is a good option in conjunction with other hot water production methods, like solar hot water heating.**
- **If possible, place a freezer/refrigerator next to your water heater to help equalize the air.**

Instantaneous Water Heater

- **Instantaneous water heaters work well with a ground source system or a solar water heating system.**
- **Make sure that the water heater is modulating according to demand.**

HAVING AN INTERGRATED SYSTEM

In the first section of this book, we learned how to have a more efficient home so you can get the greatest benefit from renewable energy. Having knowledge about efficient equipment and systems will save you a tremendous amount of money over time. With lower energy requirements, a smaller renewable energy system is needed and will therefore cost less to purchase.

We wanted to give you a bird's eye view of all the systems and functions we have discussed so far. The schematic on the next page shows a solar water heater, ground source heat pump, and an instantaneous water heater, all providing hot water to home systems very efficiently. Integrating these systems can serve your needs very effectively. Although we have not discussed some equipment shown in the diagram, we wanted you to see where it could be located and implemented so you could ask for it someday if you choose. Let's discuss the following integrated equipment and their functions one step at a time.

1. A 120-gallon hot water tank with two isolated heat transfer coils provides a large thermal mass of hot water with the ability to receive or transfer heat through these coils.
2. A cold-water storage tank provides chilled water for cooling the home.
3. A two-stage ground source heat pump with a desuperheater provides all the hot water; during the heating cycle, it transfers heat from the constant temperature of the earth to the 120-gallon hot water tank. In the cooling cycle, the heat pump reverses the cycle and chills the cold-water storage tank for air conditioning while the desuperheater assists in heating the 120-gallon tank.
4. The pump pack moves heat through fluid in pipes to and from the earth. This is accomplished by pumping this fluid through a heat exchanger located inside the ground source heat pump. In cold climates, an antifreeze fluid is pumped through the pipe loops in the earth.
5. Air handler units with ECM fan blowers on three floors receive this hot or cold water piped to coils located in each unit. Air is then blown over these coils to condition each floor as called for by thermostats.
6. An HRV (heat recovery ventilator) is installed on one or more return air ducts to bring in fresh outside air to replace stale inside air.
7. Electronic air filters clear the home of airborne dust. We would also encourage you to install UV (ultraviolet) lights to get rid of airborne pollen, germs, viruses, and any unpleasant microscopic animals you don't want to ingest.

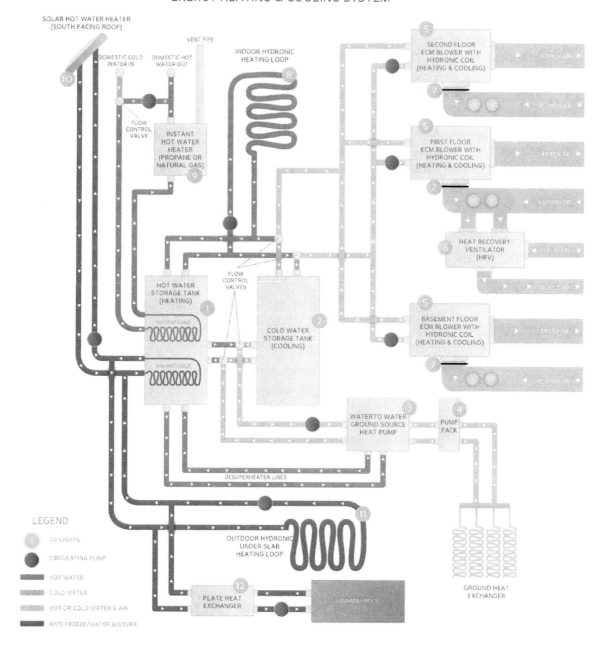

Figure 1.45

WHOLE HOUSE HYBRID RENEWABLE
ENERGY HEATING & COOLING SYSTEM

8. Hot water from the 120-gallon storage tank is circulated through radiant tubing placed under the floors to warm them.
9. A modulating 200,000 BTU instantaneous propane or natural gas hot water heater is represented for two reasons. First, when there is a very high demand for hot water,

the heat transfer coils in the tank are preheating the fresh domestic water loop to the heater, so nobody has to go without hot water. Secondly, a circulating pump past the instantaneous water heater circulates heated water back through the coils raising the temperature of the water in the tank. If demands placed on the ground source heat pump are too great, this serves as a backup heat source.

10. A solar water heater utilizes the sun for water heating, giving the ground source system a rest during those sunny days. This system has antifreeze installed in the loop and is kept safely independent from domestic water.

11. A heating loop with antifreeze is under the outside concrete to melt snow during the winter.

12. A plate heat exchanger is used to warm a swimming pool or spa.

This integrated system is all about transferring heat. This heat is transferred to and from the earth with a GSHP through a heat exchange process. Solar hot water is an excellent way to transfer heat from the sun. Some fuel heating is needed for assistance during high use times and for backup heat, but this should be an exception if your system is designed properly. The efficiency of your system is dependent on the type of equipment you choose and the design/installation processes.

The Second Step: <u>GENERATING ELECTRICITY</u>

POWERFUL CHOICES, EMPOWERING SOLUTIONS

You own your electrical appliances – why not own some or all of the electricity that powers them? For years, we have heard whispers of powerful individuals and companies buying up inventions and products that threaten their standing in the energy market. Whether that is true or not, a massive explosion of energy technology is tipping the power scale in favor of the average person.

While these new and exciting technologies are up and coming, as for now, more electrical independence is being found in our ownership of the sun and wind. These natural and free resources are ours to use at will, if we plan and empower ourselves with knowledge about their use. Look ahead and plan for your independence from utility companies with great hope.

In this Second Step, Power Generation section, we will literally empower you with an understanding of electricity, its storage, and electricity generators.

Electricity 101 – AC/DC, Volts, Amps, Watts – explained in layman's terms

Storing Electricity – Battery Basics and Net Metering

Solar Power – Solar Power Basics and Optimization

Wind Power – Wind Power Basics and Choosing a Turbine

Micro-hydro – Assessing Your Site and System Components

Emergency & Power Assist Generators – Auto and Manual Systems

Tying It All Together – An Integrated Power System at a Glance

The Right Design Saves You Money – Is a Professional Design Worth it?

The following drawing shows a range of options that use renewable energy, including several types of electricity-generating devices, a solar water heater, and a ground source heat exchanger. It is very important for you to have a basic knowledge of how they can be integrated and work best for you. We will be discussing electricity generators (devices that generate electricity) in the next section of this book. With a general working knowledge and a proper design, you will be on your way.

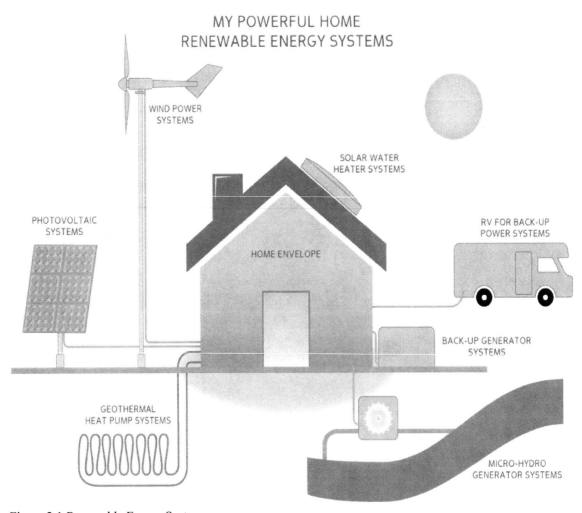

Figure 2.1 Renewable Energy Systems

ELECTRICITY 101
AC/DC, Volts, Amps, Watts
Explained in Layman's Terms

<u>ELECTRICTY 101</u>

Your Questions:

- **What are the purposes for AC and DC electricity and how are they used?**
- **What is the purpose of transformers and inverters?**
- **Can you explain amps, volts, watts and their purposes?**
- **How are the monthly watts used and calculated into my power bill?**

Figure 2.2 Commercial electrical lines.

We hope to help you understand the purpose and use of electricity so you can incorporate the terminology and the equipment in your systems. In this chapter, we will discuss in layman's terms *alternating current (AC) and transformers, direct current (DC) and inverters, and then volts, amps, and watts.*

ALTERNATING CURRENT (AC)

Alternating current (AC) is a form of electricity delivered to businesses and residences for use to operate all types of electrical devices and appliances. With AC, the flow of electricity goes both ways (alternates) to and from the generating source at approximately the speed of light. AC makes it possible to build electric generators, motors, and power distribution systems that are far more efficient than direct current (DC), and so we find AC used predominately across the world in high-power applications. High voltage transmission lines deliver power from electricity generating plants over long distances using alternating current.

Just like light and sound, all AC electricity moves in a waveform. The usual waveform of an AC power circuit is a sine wave. When you are generating AC electricity and feeding it into another electrical generating source, you must match the sine wave or you will damage equipment. It is as if two people are singing in disharmony, except sparks and heat are created. If you produce excess electricity and want to sell it back to your power company (this is called net metering), they will require that you match their form of electrical transmission with synchronizing sine wave AC electricity. Specialized equipment (a sine wave inverter) is installed to match the utility AC sine wave.

People ask us why they can't just connect their fuel generator to their home electrical panel. First, they must synchronize the generator's form of electricity to match the home's form of electricity with special equipment. Second, if the main breaker is not switched off in a power outage, electricity will be fed back into the power grid and potentially electrocute line workers.

TRANSFORMERS

Transformers change or transform AC and DC electricity from high to low voltage or from low to high voltage for different purposes. For example, a transformer could reduce voltage to a home beginning at a high of 7200 volts to a low of 120 volts. When AC electricity flows from a power generator like a dam or nuclear or coal-fired power plant, it is transformed to a very high voltage. This voltage can be as high as 155,000 to over 750,000 volts, so it can travel long distances on transmission lines.

After continually being transformed and conditioned through substations, this electrical voltage will be lowered through a transformer to match your home's voltage. Because a transformer changes high voltage to a lower voltage when coming from the power plant, it can also be changed from low to high voltage when you are generating electricity from your renewable electricity generator (e.g., solar, wind, or hydro) under a net metering agreement. The transformer doesn't care which way the generated electricity is coming from; it will always change the voltage.

DIRECT CURRENT (DC)

With DC or direct current, the electricity flows in one constant direction, giving DC its unique form of electricity delivery. DC is generated from many different sources, including wind generators, solar arrays, small hydro generators, car alternators, and some fuel generators (most fuel generators are AC). This DC electricity is generally stored in batteries and is later changed or inverted to AC before being used in the home.

INVERTERS

Most renewable electricity generators are DC and require an inverter as a component in the system. Inverters convert DC to AC electricity. It is so named because early mechanical AC to DC converters were made to work in reverse, and thus were "inverted" to convert DC to AC. An inverter is an electrical device that converts direct current (DC) to alternating current (AC); the resulting AC can be at any required voltage with an appropriate transformer.

Figure 2.3 Pure sine wave inverter.

The inverter generally changes the electricity from, 24 volts DC to 120/240 volts AC to run standard appliances and electricity needs or loads designed for this common voltage. Some electricity loss happens during this change in power (power factor); however, it is unavoidable. Be sure to purchase inverters with at least a 94% power factor, which translates to only a 6% loss of power during the change in voltage.

Some of the newest smart inverters employ a technology that keeps a weak or faulty solar panel from degrading the electrical output of your solar electric system. These smart inverters track and report on individual module/panel electricity output instantly and over time. They also alert you when a solar panel is faulty or compromised. They optimize the entire solar electric system output up to 30%. These new inverters sense and match your utility power for synchronization and they disconnect solar panel electricity output into the grid during a utility power outage to meet a utility and national electrical code requirement. Besides providing all of the above mentioned functions, they track the total wattage output of your solar array, wind turbine or other electricity generators, which can help you keep your power company honest in a net metering agreement.

At this point in technological innovation there are two types of inverters; Whole system and Micro inverters.

- Whole system inverters are wired after your solar panels, wind turbine, or micro-hydro generator, before connecting them to your home electrical system.

- Micro inverters are wired to each solar panel making them modular and simple to install. This allows you to install a few panels this year and add more next year or as desired without the expense of upgrading your whole system inverter to meet the added electrical output.

The chart (Figure 2.4) on the following page shows common applications for devices using AC or DC electricity. It is very helpful in understanding what appliances and devices generally use and why inverters and converters are needed.

UNDERSTANDING VOLTS, AMPS & WATTS

Both AC and DC conduct electricity using volts, amps, and watts. We all use these terms when discussing electrical devices and appliances, whether AC or DC, but most people do not understand their use. We hope this explanation helps your understanding.

Volts are the electrical potential or ability to deliver the flow (i.e., the pressure).

Amps are the electrical current unit (i.e., flow).

Watts are the electrical power units, and is the total of amps x volts or the actual flow x pressure (i.e., how much flow is actually used when it is delivered).

Electricity Example

Using a one megawatt generator (an electrical generator capable of delivering one million watts of electricity), let's compare this electrical generator to a waterway system; Let's say a reservoir is kept full with eight hundred million gallons of water that has a certain amount of pressure behind the dam, with a potential to deliver water downstream (volts).

Device	Alternating Current (AC)	Direct Current (DC)
Transformers	Modify AC voltage, high to low and low to high	Modify DC voltage, high to low and low to high
Converters	Change AC to DC	Change AC to DC
Inverters	Change DC to AC	Change DC to AC
Renewable Energy Generators: Solar, Wind, Micro-hydro, etc.	If the energy generating device is far away from the home, electricity must be conducted over a long distance; an AC generator can be purchased.	Most are DC. Solar is always DC.
Fuel Generators (Back-up/ Emergency): Gas, Diesel, Propane, Natural gas	AC generators are most commonly used. Most emergency generators are AC.	Some are DC. Best for off-the-grid systems to keep batteries at a full charge.
Batteries	None are AC	All are DC
Home Appliances: Freezer, Refrigerator, T.V., Micro-wave, Coffee Maker, Desktop Computer	Most are AC	You can purchase some that are DC, generally for RV or off-the-grid use.
Laptop Computer, Cell phone	Plugs into AC but is DC	Laptops run off of batteries (DC) using converters so you can plug into an outlet (AC).
Appliances: Hair Dryer, Curling Iron, Clothes Iron	Most are AC	Some are DC, generally for RV or off-the-grid use.
Lighting	Most are AC	Some are DC, generally for RV, specialty home lighting systems, or off-the-grid use.

Figure 2.4 AC DC Chart

One hundred thousand gallons of water a minute is flowing downstream by a river to water some fields below (amps).

One of the fields below needs ten thousand gallons of water an hour (watts).

The water's potential is reduced to only deliver a constant flow of 10,000 gallons an hour from the river by a flow-reducing valve (a transformer).

When the flow rate (amps x volts) is reduced by the valve, the amount of flow actually delivered (watts) is ten thousand gallons an hour.

The farmer is charged by the hour, so his watering bill reflects ten thousand gallons for one hour 10kgph (k = 1000, gph = gallons per hour).

Let's compare this to your electric bill, where 10kWh = 10,000 watts used per hour. If you are charged $0.08 per kWh, you would be charged $0.08 x 10 kWh = $0.80 per hour x 24 hours = $19.20 per day. Most homes use approximately 36kWh per day; that's 36 x $0.08 = $2.88 per day x 30 days/mo = $86.40 per month.

Because this book is about how to apply renewable energy systems to your home, let's use the same 36 kWh that cost $86.40 per month from the power company. In order to produce 36 kWh from a renewable energy system, you would need to generate an average of 1.5 kW per hour. Now let's multiply that by 24 hours in a day, 1.5 kWh x 24 = 36 kWh, and you will have produced sufficient energy for your needs. However, the sun doesn't shine 24 hours a day and wind is uniquely variable at every location. So, no matter how you plan to produce electricity, understanding how it works is important to optimizing its use.

Important Points:

- **Understanding how electricity flows and is consumed helps you manage its use.**

- **Knowing how your utility company charges you in monthly kilowatt-hours (kWh), will help you understand your bill and how much energy you need to produce.**

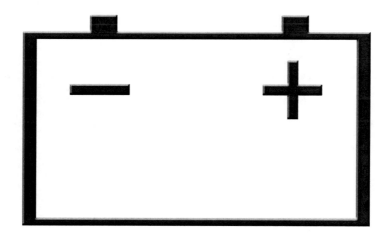

STORING ELECTRICITY
Battery Basics & Net Metering

STORING ELECTRICITY

Your Questions:

- **Why do I need batteries to store electricity?**
- **What are the most commonly used batteries in renewable energy systems?**
- **What size of battery bank do I need?**
- **How long do batteries last and how do I properly charge them?**
- **What is net metering?**

A battery is an electrical storage device that stores energy for later use. It is an electrochemical device that converts chemical energy into DC electricity, by use of a galvanic cell. A galvanic cell is a fairly simple device consisting of two electrodes (an anode and a cathode) and an electrolyte solution. Batteries consist of one or more galvanic cells. Batteries do not *make* electricity; they *store* it. All batteries store DC electricity.

Figure 2.5 Deep cycle battery.

When choosing batteries for your renewable energy system, it is best to have an expert design a system for you. This can be a complex process to get right. To give you some understanding of how they are used in a system, we will discuss: *The Need for Storing Electricity, Battery Types, Battery Bank Size, Battery Life, Battery Charging, and Net Metering.*

WHY STORE ELECTRICITY?

Let's compare storing electricity to a water reservoir system. A river or stream supplying a reservoir has a certain capacity and flow rate, like electrical generators (When discus-sing electrical generators, we are including all the renewable power generation equipment like solar, wind, hydro, etc.). This water flowing from rivers and streams is held in a reservoir, which serves as a buffer or storage for water supply needs downstream. Storage is needed because sometimes there are extremely high demands downstream that require more water than the flow rate of the rivers and streams could supply. In addition, there are times when there is little water needed downstream so the reservoir fills to capacity, like electricity in a battery.

When a reservoir is designed properly, all the watering needs, maximum and minimum, along with the projected future needs are taken into consideration. This same principal should be used when sizing battery capacity for storing electricity for your home. Generally, you wouldn't want to size your renewable electrical production equipment, solar array, or wind turbine to handle high-use times, as it would be too expensive in most cases. A battery bank should be designed to handle these high-demand times and factor in downtime when your generators are not producing electricity.

A reservoir discharges or releases water, and then refills or recharges the water in seasonal cycles. Batteries also discharge and recharge; this is called a 'cycle.' An example of a cycle would be discharging from 100% capacity to 20%, and then recharging back to 100%.

Note: A renewable energy system that is utility grid-tied and supplies power directly to the utility grid (net metered) needs no battery storage, unless you desire some backup electrical storage in case of a power outage. We will discuss net metering more a little later.

TYPES OF BATTERIES

There are many types of batteries, such as alkaline, lead acid, nickel metal hydride, and lithium ion. Nickel metal hydride and lithium ion batteries look promising for the future; however, at this time, their price is too high for the size needed in most systems with the exception of some small remote lighting systems. The two most common types of batteries used in systems today are *alkaline* and *lead acid*.

Alkaline batteries also have positive and negative plates, but are in a potassium hydroxide electrolyte solution. The plates are made of nickel and cadmium or nickel and iron. These batteries are very expensive and have environmental issues related to disposal; therefore, they are used less than the lead acid type.

Lead acid batteries have positive and negative plates made of lead that are mixed with other materials and submerged in an electrolyte solution of sulfuric acid. These batteries are the most widely used because they are less expensive due to the readily available materials of which they are made. The majority of systems for residential use are designed to use lead acid batteries.

Lead acid batteries are the most commonly used and the most readily available battery worldwide. There are many different sizes and designs of lead acid batteries; however, the most important issue is whether they are deep-cycle or shallow-cycle batteries.

§ Shallow-cycle batteries are used in automobiles for supplying large amounts of current (electricity) for a short time and will stand some overcharging without losing electrolytes. These batteries cannot tolerate being deeply discharged; if they are repeatedly discharged more than 20%, they will have a very short life.

§ Deep-cycle batteries are designed to be discharged by as much as 80% of their capacity, so they are the chosen battery type for renewable systems. It is important to note that even though they are designed to withstand deep cycling, these batteries will have a longer life if the cycles are shallow.

Sealed deep-cycle lead acid batteries, (gel cells and absorbed glass matt) are maintenance free. They never need water or an equalization charge. Sealed batteries require very accurate monitoring to prevent over-charge and over-discharge. Either of these conditions will drastically shorten their lives. We recommend sealed batteries with high quality charge controllers for remote, unattended power systems.

It is important to know that all lead-acid batteries fail prematurely when they are not recharged completely after each discharge. Letting a lead acid battery stay in a discharged condition for days at a time will cause a permanent loss of capacity.

BATTERY BANK SIZE

The size of your battery bank (i.e., group of interconnected batteries) depends on the storage capacity required, the maximum discharge rate, the maximum charge rate, and the minimum temperature at which the batteries will be used. When designing a system, all these factors are looked at and the one requiring the largest capacity will dictate battery size.

Figure 2.6 Battery banks.

The storage capacity of a battery, the amount of electrical energy it can hold, is usually expressed in amp-hours. If 10 amps are used in 15 hours, then 150 amp-hours will be required. Remember that watts is the electricity consumed by appliances or devices, and you multiply amps x volts to get watts. A battery bank connected to produce 24 volts at 10 amps would produce 240 watts (24V x 10A = 240 watts). A battery bank in a power system should have a sufficient amp-hour capacity to supply needed power when there is no wind, sun, or hydropower production. A lead-acid battery bank should be sized at least 20% larger than this amount. If there is a source of backup power, such as a standby DC fuel generator with a battery charger, the battery bank does not have to be sized for non-energy production times.

Batteries should be sized so they don't discharge too rapidly. A rapid discharge can create excessive heat to a battery bank and usually means that your battery bank is undersized for the appliances or devices they are powering. When battery banks are designed and sized properly, they will discharge at a slow rate over one to two days and may need to have a very large amp hour capacity, depending on what you are powering.

When charging your batteries, a quality charge controller will bring them up to full capacity at a charge rate that will give them the longest life. Battery charging can vary in voltage and amperage ranging from a trickle charge to a quick charge, as the charge controller monitors the state of charge of the batteries. Temperature has a significant effect on lead-acid batteries. For example, at 40° F, they will have 75% of rated capacity, and at 0°F, their capacity drops to 50%.

BATTERY LIFE

Battery manufacturers specify battery life in terms of the quantity of cycles. When batteries have lost 20% of their original capacity, they are considered to be at the end of their life, even though they can still be used. The lifespan of a battery will vary considerably with how often it cycles, the depth of the cycles, and the temperature at which it is maintained.

Battery life is directly related to how deep the battery is cycled each time. If a battery is discharged to 50% every day, it will last about twice as long as if it is cycled to 80% of depth of discharge (DOD). If cycled only 10% of DOD, it will last about 5 times as long as one cycled to 50%. For example, a battery used in a system that has experienced a consistent 25% of DOD rate would be expected to last 4,000 cycles. By contrast, a battery cycled to a constant 80% DOD would last 1,500 cycles. If one cycle equaled one day, the shallowly cycled batteries would last for 10.95 years, while the deeply cycled batteries would last for only 4.12 years. This is just an estimate. Some batteries are designed to be cycled more than once each day. In addition, batteries degrade over time, affecting their life expectancy.

Most manufacturers rate battery life cycles and capacity at 77°F (25°C) for optimum use. Even though battery *capacity* decreases at lower temperatures, battery *life* increases. Likewise, the higher the battery temperature, the shorter the life of the battery. Most manufacturers say there is a 50% loss in battery life for every 15°F increase over the standard 77°F cell temperature. As far as capacity versus battery life, this tends to even out in most systems, as they will spend part of their lives at higher temperatures, and part at lower temperatures.

Important note: Adding old batteries or dissimilar batteries to your existing system will degrade the life and capacity of your whole battery bank. Make sure you follow the manufacturer's recommendations for your batteries. If you decide to have a battery system, make sure you understand how to maintain them and they will last for many years.

BATTERY CHARGING

More batteries are damaged by bad charging techniques than all other causes combined. In a renewable energy system, the battery bank charger is called a charge controller. A charge controller has three key functions: getting the charge into the battery (charging), optimizing the charging rate (stabilizing), and knowing when to stop (terminating).

- *Charging*

  A battery charge controller receives electricity from your solar, wind, hydro, or DC fuel generators and inputs this electricity into your battery bank. A battery charge controller regulates voltage output received from electrical generating equipment to what the battery needs at the time. This voltage will vary depending on the state of charge of the battery, the type of battery, the mode of the controller, and temperature.

- *Stabilizing*

  In the initial charging stage, batteries begin charging rapidly and then the charging process stabilizes. The battery bank is then fed a constant voltage for a period of time depending on the need of the battery bank. You do not need to worry about this process, because a quality charge controller will know how long this charge time will need to last.

- *Terminating*

  When your batteries are fully charged, your charge controller will detect this and end charging or feeding electricity to your battery bank. Overcharging your batteries will drastically reduce their life. When batteries are fully charged and sit for a period of time, they will slowly discharge. A quality charger will trickle charge periodically to maintain a healthy, full charge.

Charge controllers come with different features, capacities, and efficiency ratings. Most controllers come with some kind of indicator, either a simple LED, a series of LEDs, or digital meters. Some newer models now have built-in computer interfaces for monitoring and control. The simplest controllers usually have only a couple of small LED lamps, which show that you have power and that you are getting some kind of charge. Most chargers with meters will show both the voltage and the current coming from electricity generating devices, as well as the battery voltage and charge state. Some also show how much current or electricity is being pulled from the batteries by the home.

Charge controllers range from 4.5 amp control up to 60 to 80 amps. Often, if currents over 60 amps are required, two or more 40 to 80-amp controller units are wired in parallel. The most common controls used for all battery-based systems are in the 4 to 60-amp range; however, some new controllers go up to 80 amps.

NET METERING (storage in the form of utility payment or credit)

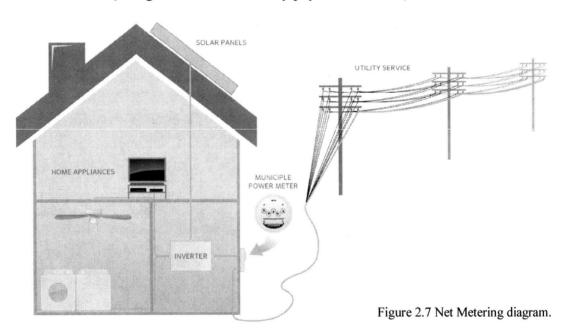

Figure 2.7 Net Metering diagram.

Net metering can be compared to storing electricity in batteries in that excess electricity generated can be 'sold' back to the power company in the form of a credit on your bill. This is far cheaper than a battery bank purchase and maintenance. In this part of the chapter on storing electricity, we will discuss: *Net metering, Interconnection, Net metering benefits, and Current concerns.*

Net metering is a special metering and billing agreement between utility companies and their customers (as overseen by the state), which facilitates the connection of small, renewable energy-generating systems to the power grid. When you are producing more electricity than you can store or use, you can transfer it to the electrical grid. There are two general ways that you can be 'credited' for excess electricity production:

1. The electricity meter that the utility company reads actually runs backward when you have excess energy that you put out onto the grid and runs forward normally when you need more than your renewable energy system is producing. You are then only billed for the 'net' amount metered.

2. Some states require that you have two meters – one that meters the electricity that you use and another that measures the amount that you have in excess to 'sell' to the power company. This is detailed on your bill and allows you to accurately measure your excess electricity production.

Power companies and the states have agreed upon how much electricity will be accepted and when they will accept it. This process should be regulated by state policy; however, not every state in the U.S. has net metering policies. As of October 2009, 42 states have net metering policies in place.

The Interconnection Standards are the rules that cover the technical requirements and the legal procedures involved when a customer electricity-generation system interfaces with the electricity grid. Generally, the local utility company must study and approve a proposed system within a framework established by the state's public utilities commission. Every state should have a set of standards in place to guide the process of connection; however, as of May 2009, only 35 states have interconnection standard policies.

The benefits of a renewable energy system that is grid-tied and net metered are many. It allows a homeowner that is installing a system in stages, as they have funds available, to have enough electricity available at all times as they ramp up to full production. Electricity credits can be collected to use in high-demand times. When a system is in a low production time, you can draw from the grid for your needs. If you don't want to deal with the cost and maintenance of a battery system, you can use your power company as your 'storage site.' There is a certain amount of satisfaction when your electrical company owes you money in the form of electricity!

The current concerns of net metering are also many; however, every year shows improvement in the overall states' policies. Wherever the standard is unclear, or where redundant or unnecessary tests or steps are piled on the existing national standards, the results can be costly and significantly affect residential energy producers. Utility companies traditionally have had the authority to decide how many systems may connect to the grid and under what circumstances. This arrangement presents a conflict of interest that can result in significant barriers to the customer.

Having pointed out the problems in the net metering and interconnection process, let's look at it in a more positive light. The following charts show the current grades given to each state for net metering and interconnection policy.

This is a yearly report produced by the Network for New Energy Choices. This represents the final tally and is illuminating; however, the whole report is worth reading because it compares each state's progress over the previous year. There are more 'improved' and 'passing' grades than there are 'failing' grades as were reflected in the 2008 report.

STATES WITHOUT STATEWIDE NET METERING		
Alabama	Mississippi	Tennessee
Alaska	South Carolina*	Texas*
Idaho*	South Dakota	
STATES WITHOUT STATEWIDE INTERCONNECTION STANDARDS		
Alabama	Mississippi	Tennessee
Alaska	North Dakota	West Virginia
Idaho	Oklahoma	
Maine	Rhode Island	

* Voluntary net metering available.

Figure 2.8 States without statewide net metering.

STATES WITH NET METERING - 2009 REPORT					
State	Net metering	Inter-connection	State	Net metering	Inter-connection
Alabama	n/a	n/a	Montana	C	F
Alaska	n/a	n/a	Nebraska	B	F
Arizona	A	C	Nevada	B	B
Arkansas	C	F	New Hampshire	C	C
California	A	B	New Jersey	A	B
Colorado	A	B	New Mexico	B	B
Connecticut	A	D	New York	D	B
D.C.	B	B	North Carolina	D	B
Delaware	A	D	North Dakota	D	n/a
Florida	A	C	Ohio	B	C
Georgia	F	F	Oklahoma	D	n/a
Hawaii	C	F	Oregon	A	B
Idaho	F	n/a	Pennsylvania	A	B
Illinois	B	B	Rhode Island	B	n/a
Indiana	F	D	South Carolina	n/a	F
Iowa	C	F	South Dakota	n/a	B
Kansas	B	F	Tennessee	n/a	n/a
Kentucky	B	F	Texas	n/a	D
Louisiana	B	F	Utah	A	F
Maine	B	n/a	Vermont	B	C
Maryland	A	B	Virginia	B	A
Massachusetts	B	B	Washington	C	D
Michigan	B	C	West Virginia	D	n/a
Minnesota	C	F	Wisconsin	D	D
Mississippi	n/a	n/a	Wyoming	B	F
Missouri	C	F	-		

Figure 2.9 2009 State Net Metering Report

Important Points:

- **All batteries store DC electricity.**
- **Most renewable energy system batteries are deep-cycle lead acid batteries.**
- **NEVER use incompatible or unequally sized and aged batteries in your system!**
- **Where possible, net metering is preferred over battery bank storage.**

SOLAR POWER
Photovoltaic Basics & Optimization

SOLAR POWER (photovoltaics or PV)

Your Questions:

- **What types of solar panels are available and how are they made?**

- **How can I get the most efficiency out of my solar panels?**

- **How can I get completely off the grid with solar panels?**

Solar power or energy is simply using the sun's rays to produce electricity or to heat water. Since we have already discussed solar water heating, we will now focus on harnessing the sun to produce electricity. Photovoltaics (PV) is the process by which the sun's rays are turned into DC electricity. This electricity can be used to power buildings. Today, most photovoltaic modules

Figure 2.10 Residential solar array.

or solar panels use silicone as their major component. Silicone cells manufactured from one ton of sand can produce as much electricity as burning 500,000 tons of coal. This clean, reliable source of electrical energy is regarded as the future of energy production.

Solar electricity is becoming more attractive and affordable as there are advancements in technology, government assistance, and as demand helps to lower solar system costs. In this chapter, we will discuss: *Planning Your PV System, Photovoltaic basics, Optimizing the Rays,* and *Off the Grid with Solar.*

PLANNING YOUR PV SYSTEM

A cost-effective use of photovoltaics and any other renewable energy system requires, first of all, a high-efficiency approach to energy consumption. As we discussed in the section, Knowing Your Home, you may need to replace light bulbs and your old appliances with Energy Star rated high efficient lighting and appliances, seal and insulate your home envelope, and install the best HVAC system.

When beginning the process of installing any renewable energy system, you must determine your energy production goal. Will you be integrating solar and wind or maybe utilizing micro-hydro? What percentage of your electricity needs do you want to produce – 50%, 80%? If you want to produce enough electricity to supply 100% of your needs, then you will need to decide to be grid-tied or off the grid entirely. Depending on where you live, you may be able to 'sell' back excess electricity produced to your utility company. (For more on this subject, see Net-metering in the chapter on Storing Electricity.)

Figure 2.11 Pole mounted so-

Once you determine your energy production goal and whether you will be integrating other electricity generating devices, then you will need to have a system designed for installation (complete or in phases depending on your available funds). A professionally designed system should alert you to the tax credits for solar power as well as the local utility company rebates. A design will save you money in the short term by allowing you to purchase only what you need. It will also save you money in the long term because a well-designed system will produce efficiently, require low maintenance, and meet your needs for 35 plus years.

PHOTOVOLTAIC BASICS

Solar cells are made of semiconductor materials, such as silicone. For solar cells, a thin semiconductor wafer is specially treated to form an electric field, positive on one side and negative on the other. When light energy strikes the solar cell, electrons are knocked loose from the atoms in the semiconductor material. Electrical conductors are attached to the positive and negative sides and the electrons can be captured in the form of an electric current or electricity. This power technology is under constant testing for efficiency at the National Renewable Energy Laboratory (NREL).

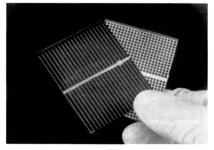

Figure 2.12 Silicone cells.

A number of solar cells electrically connected to each other and mounted in a support structure or frame is called a photovoltaic module/panel. Modules are designed to supply electricity at a certain voltage, such as a common 12 or 24-volt system. The electricity produced is directly dependent on how much light strikes the module.

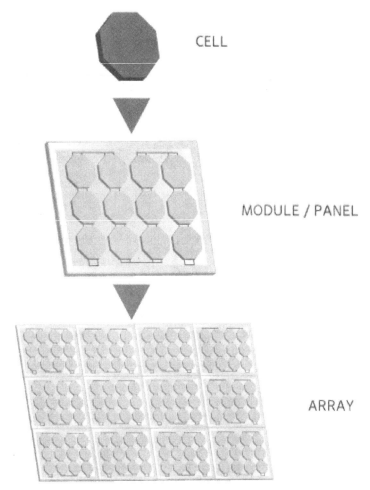

CELL

MODULE / PANEL

ARRAY

Figure 2.13 Structure of a solar array.

Multiple modules can be wired together to form an array. The larger the area of a module/panel or array, the more electricity that will be produced. Photovoltaic modules and arrays produce direct-current (DC) electricity. They can be connected in both series and parallel arrangements to produce many chosen voltage and current combinations.

Typically, a solar panel consists of 32 to 36 cells electrically connected in a series producing a panel with 15-22 volts. The silicone material used in the panel comes in three basic forms: *monocrystalline silicone, polycrystalline silicone,* and *amorphous silicone.*

MONOCRYSTALLINE

Monocrystalline silicone is produced as one large crystal and then cut into thin slices to form the individual cells. Panels made this way can be more efficient, but can be more expensive to manufacture. Monocrystalline panels are generally darker in color making them less obvious on darker roofing material, especially when the framing and racking is also dark. The panel size is often smaller because of the increased efficiency.

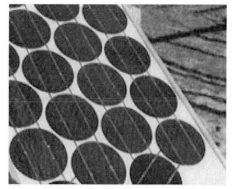
Figure 2.14 Monocrystalline solar panel.

POLYCRYSTALLINE

Polycrystalline silicone is generally cast in blocks and the final cut slices consist of many smaller crystals. Manufacturing costs can be lower; therefore, these panels may be a little cheaper to purchase. The efficiency is a little lower and they generally do not perform as well as monocrystalline types at higher panel temperatures. However, new manufacturing practices are bringing monocrystalline and polycrystalline much closer to the same price range and efficiency.

Figure 2.15 Polycrystalline solar panel.

AMORPHOUS

Amorphous silicone panels and flexible units are produced by a completely different and cheaper process by depositing vaporized silicone directly onto a backing material. This results in a cheaper panel; however, the efficiency is about half that of mono or polycrystalline types. This means you need twice the panel surface area to get the same output in power. The advantage is that amorphous can be applied to flexible backing and has the ability to be laid on uneven or curved

Figure 2.16 Amorphous solar panel

surfaces. However, having said that, this technology is in a steep improvement period.

OPTIMIZING THE RAYS

Solar panels are available in a variety of sizes. The larger the panel size, the greater the electrical output should be. Common sizes range from 10 watts to 300 watts. However, any two panels of the same output from different manufactures may have different efficiency ratings. It is important to understand what an efficiency rating is. It is very interesting how little of the sun's rays are actually captured and become electricity for your home use.

Figure 2.17 Sun tracking pole mounted solar arrays.

The efficiency rating of solar panels is the percentage of power that is converted to electricity from the sun's energy. Scientists believe that the maximum efficiency or percentage of solar energy that any silicone cell will produce is 40%, although most PV panels only range from 13%-18% efficiency.

The maximum output will only be achieved when the panel is pointing directly at the sun and the panel temperature is 25°C/77°F or less. A 100 watt-rated panel will only produce 100 watts in a given temperature range, at a 90° angle to the sun, and with no shading. While you cannot control outdoor temperatures, you can control the angle between your panels and the sun.

A solid mounted panel on your house will vary in efficiency and power output as the sun moves across the sky because it is not at that ideal 90° angle to the sun most of the day. As you can see from the following table, if you have stationary mounted panels, the output will vary depending on the time of day. These figures are only approximate and depend on the type of panel. Also remember that during the winter months the sun is much lower in the sky (lower angle to the panel) and the days are much shorter resulting in approximately half as much electricity collected per day.

Angle of Sun to Panel	% of Rated Output (Approximate only)
90 Deg	100%
75 Deg	95%
45 Deg	75%
30 Deg	50%

Figure 2.18 Angle of sun to panel versus output.

You are not subject to these output figures if you install some kind of sun tracking device. For this reason, power companies like to use automatic sun tracking systems to take full advantage of the suns power all day long. A tracker starts in the morning facing east and automatically moves with the sun until the sunsets, then it resets to the east for sunrise. It is just like a sunflower. There are You Tube videos that demonstrate sun tracking devices in action.

However, the cost and maintenance of a sun tracking system for residential use makes it less attractive. Some would rather install a few more solid mount or fixed pole mounted panels to increase output than pay the price of a sun tracking system and its maintenance. After all, solar panels are non-mechanical and adding a tracker makes them mechanical. In the end, it is just another choice you will have to make.

What about shaded conditions? When it comes to solar energy, the power of the solar array is only as strong as the weakest performing panel. Therefore, in real world conditions, where shade is common from sources like trees, chimneys, power lines, and even bird droppings, solar efficiency can be decreased significantly. In fact, as little as 10% shading can result in as much as 50% lost power. Power optimizers and some inverters maximize the energy potential of each individual panel so as much as 50% of the lost energy can be reclaimed. You don't have to be living in the desert to have an efficient solar system installed.

Power optimizers and some inverters can help make solar power a reality by maximizing the solar energy harvest and increasing kilowatt-hours in real world conditions. One of the companies offering a solar power optimizer we recommend, Solarmagic.com/real_world. (Also see the smart inverter information in Electricity 101 chapter)

Note of caution: Be careful when buying solar panels at greatly reduced prices. Ask these questions: What is the Efficiency Rating? Does it have a warranty? Is it a low-grade panel? In addition, when choosing a company to install your PV system, ask: Is my installer an electrical contractor or NABCEP certified?

OFF THE GRID WITH SOLAR

If you are planning to be totally off the grid with your solar array, there are three important things to consider:

1. Properly sized backup generator (See the chapter on emergency and power assist generators for the details)

2. The right sized battery bank (See the chapter on storing electricity for details)

3. A large enough solar array (An off the grid solar system can be expensive, so a proper design is absolutely necessary. It is very important that you downsize your electrical consumption by installing extremely efficient lighting and appliances as we recommended in earlier chapters. For example, every watt of power you save by reducing electricity needs can save you $8.00 on an array cost. At $8.00 per watt, quality energy efficient devices and appliances become relatively inexpensive.)

Let's say that you caught the right vision and reduced your electricity consumption to well below the average household. For example, a good number to shoot for would be half the national average at 18 kW (18000 watts) per day. For you to store at least 18 kW per day, and your area had good sun for 6 hours a day, you would need to generate at least 18kW in

6 hours. 18kW/6 hours of sunshine = 3 kW. At $8.00 per watt for an array, this solar electrical system would cost you $24,000.00 to install. You would then have a battery bank and a backup fuel generator for non-producing days that are sized to meet the total needs of the home as well.

The illustration below, figure 2.19, depicts the basic components of an off the grid solar system.

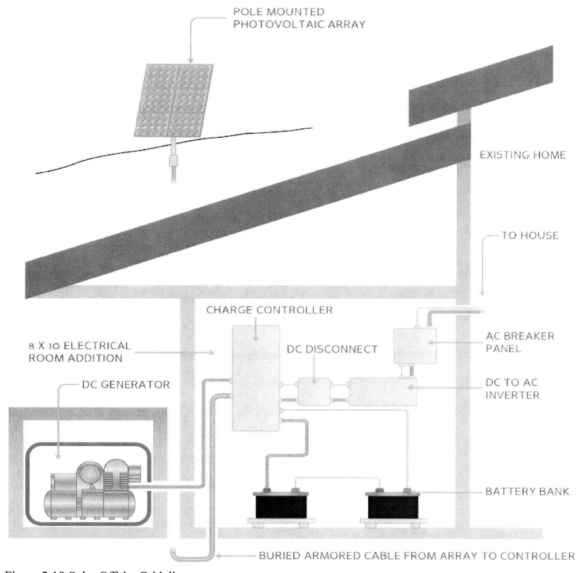

Figure 2.19 Solar Off the Grid diagram

ON THE HORIZON

Nanotechnology in solar is up and coming with all kinds of possible applications from film to paint. We are just beginning to see what the next generation of PV can offer in terms of application and efficiency. One company has achieved an efficiency rating of 16.4%. That is as high an efficiency rating as some panel manufacturers. The Idaho National Laboratory claims the highest possible efficiency rating for film PV, but is not quite ready to send it to commercial production. Keep your eye on the horizon for this technology's advancement, as it could drastically change the solar industry.

Important Points:

- **Most PV panels range from 13% to 18% efficiency.**

- **The larger the panel size, the greater the electrical output.**

- **Maximum output will only be achieved when the panel is at 90° angle to the sun and in a temperature of 77°F or less.**

- **Most modules have 25-year manufacturer's warranties, but they can produce electricity for longer.**

WIND POWER
Wind Power Basics & Choosing a Turbine

WIND POWER

Your Questions:

- **How can I determine if a wind turbine will work on my property?**
- **What size of wind generator should I choose?**
- **How much power do I need to generate?**
- **What type and height of wind tower should I choose?**

When you own a solar array, a sunshiny day will enhance your mood and make you feel productive. Generating electricity from the wind can have the same effect. Put up a wind generator and let the wind become a power-producing friend. Instead of dreading a windy day, let it translate into more electrical power independence by utilizing this free energy.

Figure 2.20 Residential wind power.

You just can't escape the fact that the amount of wind you have at your property determines how much power you can expect from a wind turbine. Though few people would consider placing a solar panel in the shade and expect it to work, the number of people who try the equivalent with a wind turbine is surprising. To better understand wind energy , in this chapter we will be discussing: *Wind Basics, Wind Generator sizing, Choosing a Wind Turbine, and Wind turbine Tower and Height.*

WIND BASICS

The power in the wind is a function of *air density*, the *blade area intercepting the wind,* and the *wind speed.* Increase any one of these factors and you increase the power available from the wind.

Air density varies with temperature and elevation. At sea level, the air density is 100% and this decreases to about 92% at 3000'. Warm air is less dense than cold air. Any given wind turbine will produce less in the heat of the summer than it will in the dead of winter with winds of the same speed.

The blade area intercepting the wind will be determined by how much power is needed. For example, you may see blades the size of an airplane propeller turning a 500-watt wind generator or blades as high as a twelve-story building tuning a 1,000,000-watt generator, using the same wind. The larger the blade area, the more wind there is to push the blades that turn a larger generator.

Wind speed is also a function of the power in the wind. Double the speed of the wind and you increase the power available by eight times. Even a small increase in the wind speed substantially boosts the power in the wind. This is why we emphasize the importance of putting your turbine where the winds are best.

Because obstructions near the ground disrupt the flow of the wind, wind speeds typically increase with height. Sometimes, wind speeds increase dramatically with height over rough terrain. Figure 2.21 illustrates what an ideal turbine placement should look like.

Since small changes in the wind speed have such a profound effect on the power in the wind, it is important to study the wind conditions at your selected site. Some people might consider installing a small micro wind turbine (about 40 watts) at their site to see what kind of power output is achieved over time. Then, if the results are positive, they could install a larger wind turbine to expand their system. If you are not satisfied with your original results, you can sell your wind generator online or through some other means and your loss shouldn't be too bad. Renewable energy is a hot topic anywhere on the internet. It will sell.

If you would rather not buy an experimental wind turbine, you can buy a device that measures wind speed called an anemometer (Figure 2.22). It can be purchased with a recording device and put in place for a year, recording all of the seasons' wind. However, to install a

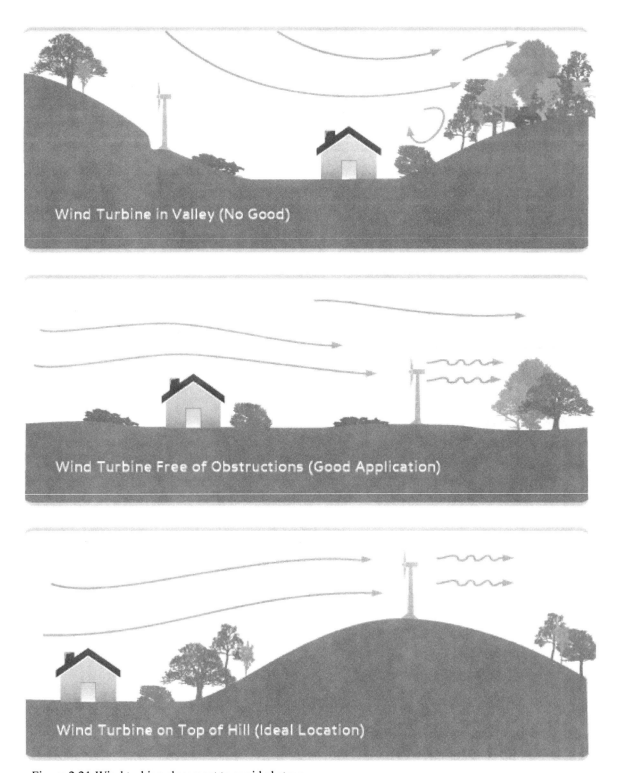

Figure 2.21 Wind turbine placement to avoid obstruc-

recording anemometer, costs about the same amount as a small wind turbine. While the wind turbine is in use, its actual energy production can be tracked to measure the wind potential. As a less costly option, some states have anemometer loan programs. To find your state program try going to: Windpoweringamerica.gov/anemometer_loans.asp

Figure 2.22 Anemometer.

If you don't want to spend money on a small turbine or anemometer, this doesn't mean that you shouldn't study the wind at your site. Go fly a kite! Attach streamers to the line every ten feet and watch how those streamers near the ground move and flap, while those higher up smooth out. The location at which the streamers fly smoothly is where you want your wind turbine. This is simplistic and does not tell you how much wind you have, but it does give you an idea of where to find your best wind.

Another suggestion is to simply look at the vegetation flagging around your home. The flagging chart (Figure 2.23) gives you an idea of what kind of wind you might have to work with; however, you shouldn't base your purchase on this alone.

FLAGGING
GIRGGS-PUTNAM INDEX OF DEFORMITY

INDEX	0 NO DEFORMITY	I BRUSHING AND SLIGHT FLAGGING	II SLIGHT FLAGGING	III MODERATE FLAGGING	IV COMPLETE FLAGGING	V PARTIAL THROWING	VI COMPLETE THROWING	VII CARPETING
WIND MPH	0-6	7-9	9-11	11-13	13-16	15-18	16-21	22+
SPEED M/S	0-2	3-4	4-5	5-6	6-7	7-8	8-9	10

PREVAILING WIND

Figure 2.23 Wind speed flagging.

To locate sources of wind data, go to Windpoweringamerica.gov/wind_maps.asp. Click on the wind map for your area and look for informative, in-depth publications on this site. Also, try the National Renewable Energy Laboratory website at www.nrel.gov/learning/, which contains a wealth of up-to-date information.

WIND GENERATOR SIZING

Basically, the sizing of a wind turbine takes into consideration three major factors (This assumes you have good constant wind available to utilize):

1. The first is *electricity consumption*, which was discussed in the chapter on the HOME ENVELOPE. This helps to reduce the size of the wind turbine designed into your system.

2. The second consideration is whether you have *other electricity-producing units* in the system, such as a solar array or back-up fuel generator.

3. The third major factor is your choice of *electricity storage,* as referred to in the Storing Electricity chapter.

To determine *electricity consumption*, let's start with your electric bill. Here is an example. On the utility bill, you will find a statement similar to "1080 kW hours usage." This is the number to look for. If there were 30 days in the month, you would divide 1080 by 30 days. Then, divide that number by 24 hours in a day to get average use. 1080 kWh/30 days = 36 and now divide 36/24 hours = 1.5 kW per hour. This is your average hourly usage (for that month).

You should do this for several months during the year to determine your true average, or at least take the highest and lowest months of the year and average them first before doing the above calculation. Unfortunately, this number does not tell you the peak usage during a given day. Generally, peak times are in the morning (breakfast time) and evening (dinnertime). In the U.S., the average home electricity use ranges between 1.5 kWh (1500 watts per hour) and 2.5 kWh (2500 watts per hour).

Incorporating *other electricity-producing units* into your renewable energy system to assist your wind generator is an important consideration. Remember that a wind turbine generator is not like a fuel-based generator that always produces its rated wattage as long as there is fuel in the tank. A wind generator varies its production with the weather. What does it actually produce? Here is an example of one type

Figure 2.24 Wind power and Solar power combined .

of wind generator advertised online. It claimed to be a 3.5 kW wind generator. When you read a little more, you find out that this output is at a maximum with a wind speed of 28mph. It only produces approximately 2 kW at a wind speed of 12mph. So, if your average wind speed is 12mph, you will average 2 kW per hour of electrical production with this 3.5 kW wind generator.

While it is possible, generally it would be inefficient (in terms of dollars per watt) to design a system to be completely off the grid with wind and battery backup alone. Unless you can prove reliable wind for several hours each day, be sure to integrate another source of electricity production. This backup power could be solar, hydro, or fuel generators, all charging your battery bank.

Assuming that you use 1.5 kW per hour of electricity, multiply that by 24 hours in a day to understand how much electricity you should be producing each day; 1.5 x 24= 36 kWh. If a wind generator produces 2 kW at an *average* wind speed of 12 mph for eight hours a day, then you would have 16 kWh generated that day (2 kW x 8 hrs = 16 kWh). You would need another 20 kWh from solar, hydro, or a fuel generator to make up the difference. In 24 hours, there will be times when you produce more than you need. This energy will need to be stored.

There are two effective ways *electricity storage* can be accomplished: First, a battery bank will act as a power reservoir, and with the proper number of batteries, it will allow your system to handle high power demands and will be available when motors need instantaneous high starting power. Secondly, a grid-tie system will automatically draw the extra power from the grid when you need it and will feed back excess electricity generated. For a greater understanding of battery storage and net metering, you can read the chapter on Storing Electricity.

CHOOSING A WIND TURBINE

There are many different and innovative wind turbines coming into the renewable energy market yearly. However, most wind turbines being sold today are *horizontal axis* turbines with two or three blades. *Vertical axis* turbines are also popular, but less efficient.

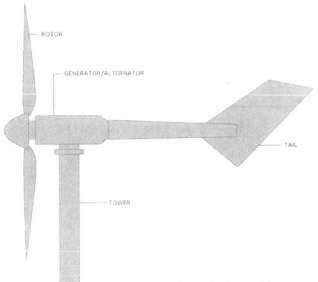

A *horizontal axis wind turbine* has rotors that spin in front of the generator and a tail that is horizontal to the horizon. The diameter of the rotor defines its "swept area," or the amount

Figure 2.25 Horizontal wind turbine parts.

of wind intercepted by the turbine. The turbine has a frame onto which a tail is attached to keep it facing into the wind. It cannot handle turbulent wind that swirls, stalls, or changes direction quickly.

A *vertical axis* wind turbine is a popular turbine in the residential market. It is visually more appealing, quieter, can be mounted on roofs or on lower poles, and it can better handle the turbulence found in urban/suburban areas. But they require more maintenance due to that same turbulent wind and have not proven to be as efficient as the horizontal axis turbine. New vertical axis designs are tested regularly at the National Renewable Energy Laboratory. We can look forward to new designs that have passed these testing rigors for use in urban areas. There are several informative articles on vertical axis wind turbines at Motherearthnews.com when you search the phrase: horizontal and vertical axis. See figure 2.26.

No matter what type of turbine you choose, they are all sold in terms of watt capacity. Micro turbines (40w-100w) are generally suitable for recreational vehicles, sailboats, or ranch electric fence charging. Mini turbines (100w-500w) are a bit larger and are best used when the wind is optimal in a shorter time frame, such as for a summer cabin with low electrical demands. So be sure to choose a residential turbine that has the ability to generate from 500w to 5 kW, depending on your needs.

Figure 2.26 Vertical axis wind turbine.

WIND TURBINE TOWER AND HEIGHT

There are many options to choose from when in the process of installing a wind generator.
Of these, the tower type and height are essential to having an optimal and efficient use of wind energy.

TOWER TYPE

Basically, there are two types of towers, *self-supporting (freestanding)* and *guyed*.

A *self-supporting or freestanding tower* is what you generally see in the large commercial turbines. You can also have this type of tower in your home application, but it must have a

Figure 2.27 Freestanding

deep and substantial base of concrete to anchor it so that it can withstand the pressure of the wind. It is usually more expensive to install because of the base required. It is attractive to some homeowners because it does not require a large space or the cabling that guyed towers use. This type of tower can be hinged at the bottom for maintenance.

A *guyed tower* assembly is the most commonly used in home wind power systems. Guyed towers, which are the least expensive, can consist of lattice sections, pipe or tubing (depending on the design), and supporting guy wires. They are easier to install than self-supporting towers. Because the guy radius must be one-half to three quarters the tower height, guyed towers require more space. A guyed tower can be hinged at the bottom for maintenance just as a freestanding tower can be.

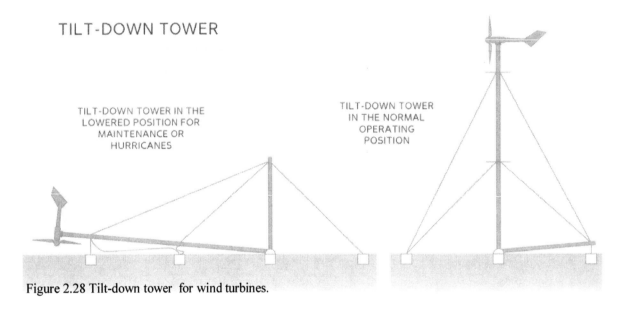

Figure 2.28 Tilt-down tower for wind turbines.

Although tilt-down towers are more expensive, they offer the owner a convenient way to perform maintenance on smaller lightweight turbines, usually 5000 watts or less. Tilt-down towers can also be lowered to the ground during extremely high winds from hurricanes and tornados.

Note: Aluminum towers are prone to cracking and should be avoided. Mounting horizontal turbines on rooftops is not recommended. All wind turbines vibrate and transmit the vibration to the structure on which they are mounted. This can not only lead to noise and vibration on the building, but rooftops can also cause excessive turbulence that can shorten the life of the turbine.

HEIGHT

As a general rule of thumb, install a wind turbine on a tower with the bottom of the rotor blades at least 30 feet (9 meters) above any obstacle within 300 feet (90meters) of the tower. For a relatively small investment, a taller tower can bring a very high rate of return in power production.

For example, raising a 5 kW (5000 watt) wind generator from a 60' tower to a 100' tower will increase the system cost by about 10%, but it can produce up to 30% more power. So install the turbine as high as you can to recuperate the installation cost more quickly.

Installing your wind turbine on a hill is ideal, and may allow you to have a shorter tower. However, in general, the higher the tower, the more power in the wind. Remember, a tower also raises the turbine above air turbulence that can exist closer to the ground. This DOE picture indicates a secondary rule of thumb, that turbine placement should be at 2 times the height of your highest obstruction.

OBSTRUCTION OF THE WIND BY A
BUILDING OF HEIGHT (H)

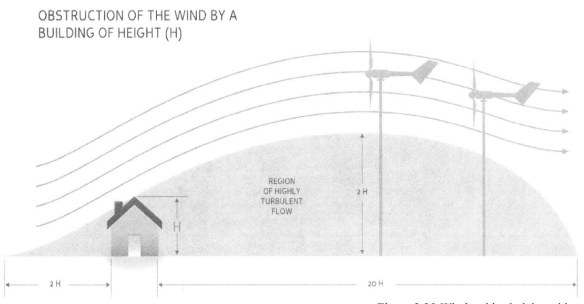

Figure 2.29 Wind turbine height guide.

Other issues to consider with wind turbines are local, municipal, and state regulations. Is there a permit process? Will the turbine be opposed by neighbors that could be bothered by the noise or an obstructed view? Before purchasing anything, be sure to have these questions answered.

As you have read through this chapter, you have seen that there is a lot to consider when choosing and installing a wind turbine. This is why we recommend that you have it professionally designed as part of an integrated energy system.

Important Points:

- **Find the best wind at your site.**
- **An integrated system is best if you want to be off the grid.**
- **Install your wind turbine as high as possible to get the best wind.**
- **Install a hinged tower for easy maintenance.**

MICRO-HYDRO POWER
Assessing Your Site & System Components

MICRO-HYDRO POWER

> ### *Your Questions:*
>
> - **Will a micro-hydro generator work at my site?**
> - **Can I develop enough water pressure to use a micro-hydro generator.**
> - **How do I determine water pressure and water flow?**
> - **How much power can I generate with a micro-hydro system?**
> - **What are the components of a system?**

Figure 2.30 Micro-hydro generator.

Micro-hydropower is making a comeback for electricity generation in homes. Even though there may be few people that have this energy production option, we decided to add it to the book for those that may not have realized that they could utilize their stream, pond, or irrigation this way. Finding a place for a hydro system is much more site-specific than a wind or solar energy system. A sufficient quantity of water must be available, with enough drop in elevation to create pressure. To give you a greater understanding of micro-hydroelectricity, we will discuss *Intro to Micro-hydro, Measuring Head Pressure, Measuring Flow, Determining Available Power, Components of a Micro-hydro System, and Considerations for Grid Tie and Off Grid.*

INTRO TO MICRO-HYDRO
Hydroelectric power is produced through the kinetic energy of falling water or water under

pressure. In the most common application, water is diverted from a river or stream, or held by a dam and then routed through a pipe where it develops pressure. This pressure is forced through a properly sized nozzle. The nozzle develops a jet stream force of concentrated water to rotate a paddle wheel connected to a turbine. The turbine is forced to rotate by the force of the water pressure and water flow. The turbine is mechanically connected to a generator, which is then forced to rotate, generating AC or DC electricity.

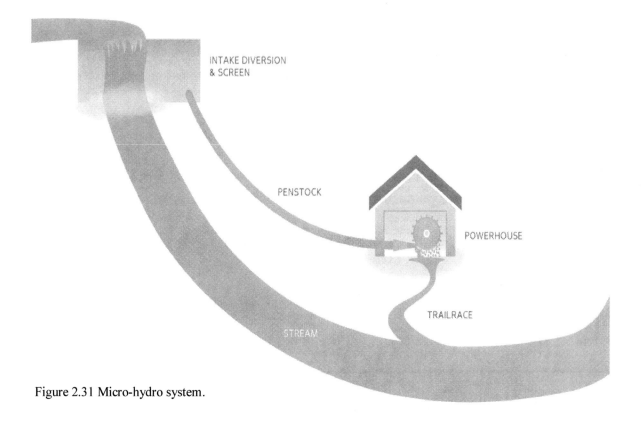

INTAKE DIVERSION & SCREEN

PENSTOCK

POWERHOUSE

TRAILRACE

STREAM

Figure 2.31 Micro-hydro system.

The amount of power available depends on the *head* pressure and the amount of water *flow*. The vertical distance the water falls is called *head* and is usually measured in feet, meters, or units of pressure (see figure 2.31). The quantity of water is called *flow* and is measured in gallons per minute (gpm), cubic feet per second (cfs), or liters per second (l/s). More head pressure is usually better because it requires less flow. The greater the distance water has to fall, the more head pressure is developed, thus requiring less water. High flow and less head can add up to the same amount of power generated as high head with low flow. The difference will be in the cost of the equipment, because high flow/low head equipment is larger and more expensive then high head/low flow equipment. To harness hydro power there must be enough head and flow to turn a turbine at enough rotations per minutes (RPM) .

Here are some simple rules for evaluating your site.

> <u>Rule #1</u> - you should have at least 6 1/2 feet (2 meters) of head for a hydro turbine.
> <u>Rule #2</u> - use a water flow you can count on year-round.

To consider a site for a potential hydro system, you must first determine the head and flow of the site. You cannot have a meaningful discussion about a hydro project without first knowing these two parameters accurately.

MEASURING HEAD PRESSURE

The head for a given site can be measured using a transit or a pressure gauge. Head pressure is the distance from the top surface of the water at the intake to the turbine. This is where

HEAD PRESSURE

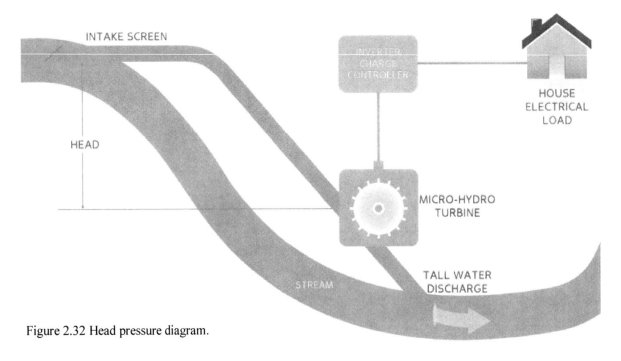

Figure 2.32 Head pressure diagram.

the power is generated. If the pipe is already in place, stop the water from flowing, and then use a pressure gauge to measure the pressure. This will relate precisely to the head at the site.

You can also measure the head pressure by using a garden hose long enough to stretch from the expected intake to the outlet. You will also need a common pressure gauge that has a range of 0 to 30 psi.

Step 1 - Fill the hose with water and close off the ends using shut-off valves.

Step 2 – Submerge one end of the hose in the water at the intake point and the other at the outlet point.

Step 3 - Open both ends of the hose, starting with the intake end. The water should begin to flow – maybe quite slowly (due to friction).

Step 4 - Allow the water to flow through the hose for several minutes – to be certain all the air is removed.

Step 5 - Now place a pressure gauge in the outlet end and seal tightly. This will stop the flow.

Figure 2.33 Pressure gauge

Step 6 - Allow the gauge to settle to a reading. Divide that number by .43 (a standard multiplier) that will give you feet of head. For this example, let's say the gauge reads 12 psi. 12 psi/.43 = 27.9' of head pressure.

MEASURING FLOW

The amount of water that you have available should be measured to find the flow. The method used to measure flow depends upon the volume of water – *high, medium or low.*

High Flows

Many larger streams/rivers are monitored, and flow data will be available at the United States Geological Survey website. If there is no data, you will need to measure the water flow or estimate the flow based upon the size of the river and its water velocity. You can also measure the flow of water that you want to divert.

Medium Flows

The weir method can be used for smaller streams or for streams with dams. In this process, water is diverted to flow over a weir with a known width. The depth of the water flowing over the weir plate is measured, and then by using a weir table, the flow is known. You can find a weir table at: Energybible.com/water_energy/measuring_flow.html

Low Flows

For a stream with low flow, a bucket may be the best approach. With this method, water is diverted so it can fill a five-gallon bucket. The empty bucket is placed under the diverted flowing water, and the time it takes to fill the bucket is recorded. Do this test several times and take an average value. For example, if a 5-gallon bucket fills in 2 seconds, the flow is 2.5 gallons per second. Multiply 2.5 gal x 60 sec =150 gallons per minute (gpm).

When considering available water flow, you should determine the minimum flow that can be reliably supplied to the turbine year round. A generator will be sized to meet the low flow rate.

DETERMINING AVAILABLE POWER

Now that you have the head pressure reading of 27.9' and the flow rate of 150 gpm, you can get an approximate idea of the power that can be generated by using the following equation:

$$\text{Head x Flow x .113} = \text{Watts}$$

Or

Head (feet) x Flow (gpm) x .113 (standard multiplier) = power (watts)

So

27.9' x 150 gpm x .113 = 473 watts

473 watts x 24 hrs = 11,352 watts (11.352 kWh)

This could amount to about half the power requirement for an efficient home. So, how much power do you use now?

A great place to start is with your utility bill. This should give you the monthly kilowatt hour usage; let's say 1080 kWh. For example, if there were 30 days in a month, you would divide 1080 by 30 (days), and then by 24 (hours per day) to get average use per hour. This equals 1.5 kW per hour.

You should do this for a year to determine your true average. At minimum, average several bills that cover the highest and lowest use periods. A typical home uses about 1080 kWh per month.

How much power do you need to generate?

If you install a system with batteries, or a grid-tie system, you can use the average daily kWh of electricity use for the minimum amount of power you should generate. A battery system will act as a power reservoir, and with the proper number of batteries, will allow your system to handle peak demands when motors are starting. A grid-tie system will automatically draw the extra power from the grid during peak usage.

Off-grid hybrid systems with solar or wind systems are another option where batteries store power for equipment with motor-starting needs, like refrigerator compressors, AC units, and pumps. The right system to fit your needs should be designed by a professional.

SYSTEM COMPONENTS
Although micro-hydro systems vary from site to site, there are some basic components that

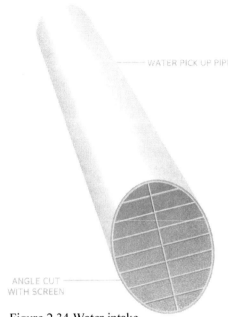

WATER PICK UP PIPE

ANGLE CUT
WITH SCREEN

Figure 2.34 Water intake

most have in common: Water Intake, Selling Tank, Pipeline, Shut Off Valve, Turbine Generator and Tailrace.

Water Intake

The water intake must be located so it will always supply the necessary and adequate amount of water to the turbine. It must include a screen to prevent debris, fish, or rocks from entering the turbine. A properly designed and constructed screen at an angle will be self-cleaning and require little maintenance. A poorly designed screen will require upkeep and will rob the system of power.

Settling Tank

A settling tank is not necessary, but is highly advisable. A concrete tank generally used in septic systems could serve as your settling tank. This serves as an area for the water from the intake to decelerate and for any fine materials, such as sand and gravel that passed through the intake screen, to settle and not flow into the turbine. As a general rule, the set-tling tank's capacity should be about 20 times the turbine's water usage in one minute. For example, if your turbine uses 100 gallons per minute, the tank should hold at least 2000 gallons of water.

Pipeline

The pipeline will deliver water to the turbine from the settling tank and must be sized properly to prevent frictional losses from robbing available power. This will be a cost/benefit trade-off since a long pipe can cost more than the turbine. Pipe should be buried for protection and to prevent

Figure 2.35 Micro-hydro water pipeline.

freezing in cold weather climates. The pipe should include a breather, ideally near the settling tank. This will prevent the pipe damaging event known as an implosion. Depending upon the pipe chosen, if the intake were to be shut off suddenly, a huge vacuum would develop in the pipe causing the pipe to implode. This is caused by the water's momentum and can be of a large enough magnitude to cause the pipe to collapse. A small vent can save you thousands of dollars.

Shut-off Valve

A shut-off valve is necessary, and should be directly in front of the turbine in case an immediate shutdown of the system is required. This valve should be high quality and very durable.

Figure 2.36 Shut off valve.

Pressure Gauge

Figure 2.37 Pressure gauge.

A pressure gauge should be installed at the end of the pipe, right in front of the turbine shut-off valve. You will find this gauge to be invaluable. First, it will be a quick way to monitor how the system and pipe are performing. A quick reading of the pressure tells you if there is a change or a problem. Second, you can monitor this gauge when you close the turbine shut-off valve. Turn it off slowly so as not to create too much pressure and burst the pipe.

Turbine Generator

The turbine will extract energy from the flowing water, and turn it into mechanical energy that turns the generator to create electrical energy. System efficiencies range between 65% and 80% depending upon the turbine style and design.

Turbines need to be hardy machines, preferably made of steel rather

Figure 2.38 Hydro turbine.

than plastic. Steer clear of turbines made from plastics unless it is just a hobby project. If your electrical needs will be dependent upon your turbine, make sure you select a machine made with steel components. Ideally, the turbine, generator, and electrical control boxes should all be housed in a weatherproof structure. It should resist bad weather, animals, and intruders.

Tailrace (water outlet)

A necessary and sometimes forgotten component in design is the tailrace or water outlet. Water must have a convenient and non-restricted path back to the stream or pond. In cold climates, the tailrace must be designed to prevent freezing in the winter. The tailrace should also be designed to prevent erosion, since a large continuous volume of water may pass through them.

Figure 2.39 Tailrace (outlet).

CONSIDERATIONS FOR GRID-TIED & OFF-GRID

Grid Tied and FERC

The mechanics of grid-tied or off-grid systems are well established. However, if you plan to do a hydroelectric grid tie in the U.S., be prepared to deal with the Federal Energy Regulatory Commission (FERC). This requirement is unique to hydroelectric generators that are grid-tied. However, if you are producing less than 5 megawatts of electricity, you only need to apply for an exemption from licensing. It should take significantly less time to get the exemption notification than the licensed permit, which can take at least one year.

Off-Grid

The off-grid solution is quite easy and the equipment is readily available. There are a couple of approaches to this method:

      1.) Produce enough power to run your home, farm, or business. This requires the largest turbine needed to generate enough power to handle peak loads (like motors starting).

2.) Charge batteries to store power for a time when it is needed. This allows a smaller turbine to be used, one that will generate sufficient power for your total use. You will need to use an inverter to provide AC Power, because batteries operate with DC voltage and current.

2.40 Battery Bank for off the grid living and micro-hydro power

You can research many on-line sites for further information. The state of Oregon has an informative micro-hydroelectric site that you might want to take a look at. Oregon.gov/ENERGY/RENEW/Hydro/ Hydro_index.shtml

---

Important Points:

- **Head pressure and water flow must be determined before choosing a micro-hydro generator.**

- **A generator must have at least 6.5' of head pressure.**

- **Size your system to the lowest flow of the year. An integrated system is best if you want to be off the grid.**

- **A professional designer or supplier should be involved in sizing your system.**

EMERGENCY BACK-UP POWER GENERATORS
Auto and Manual Systems

EMERGENCY BACK-UP POWER GENERATORS

> ### *Your Questions:*
>
> - **What size generator do I need?**
> - **Can I use my RV or pull start generator as emergency backup?**
> - **When would I need an automatic generator system?**
> - **What are the components of a system?**
> - **Should I choose and AC or DC generator system?**

Figure 2.41 The electrical grid is vulnerable to any passing storm. These photos show the results a severe freak storm that hit northern Utah in August 2006. It lasted no longer than a half hour, but included hurricane force winds exceeding 100 mph and torrents of rain. No deaths or serious injuries occurred; however, electricity was not restored for three days.

Some areas of the United States experience several power outages a year with some lasting for a week or more. Severe weather storms also contribute to extended power outages. Many Homeland Security experts continue to warn us about a potential terrorist threat to our power grid system. The truth is, that practically everyone is subject to a power loss in one way or another. Even when

Figure 2.42 Dangerous downed electrical pole.

power losses are anticipated with rolling blackouts, people can literally be stopped in their tracks. More frequent power outages are predicted in the future as our power grid continues to be loaded with more and more demands.

We believe that self-preparedness is the best way to enjoy peace of mind in the energy market. Let's talk about several options you have to empower yourself and avoid trouble when the power grid fails to deliver needed electricity. In this chapter, we will discuss *Generator Sizing, Generator System Costs, Manual Generator Systems, Automatic Standby Generator Systems, Generator Fuel Types,* and *AC & DC Generators.*

GENERATOR SIZING

When sizing a generator, it is very important to reduce electrical power consumption. As discussed earlier in this book, this can be by done by installing appliances and lighting fixtures that reduce the overall load on your electrical system.

Choose which appliances you need to power during an outage, including the furnace fan (if it is gas or propane), a refrigerator/freezer circuit, a few convenience receptacles, and lighting circuits. These will be routed from your existing panel to a new emergency panel.

Now that you have an emergency panel, you need to determine how much electricity (in watts) your emergency circuits use for sizing your generator. This would be most accurate by turning everything on that is connected to your new emergency panel and using an amp meter to see what the current draw is. Your electrician could help you with this very easily.

Figure 2.43 3250 watt AC backup generator.

Let's say, for example, that your amp draw is measured at 30 amps total and your voltage is 120V, 30 x 120 = 3600 watts total. As a general electrical rule of thumb, we would recommend that you figure 1 ½ times (150%) the 3600 watts so you're below the full demand of power required to run your basic necessities. The total power demand of 3600 x 150% (1.5) = 5400 watts. You would need to purchase a 5500 watt to 6000-watt generator to run your basic needs during an emergency. A 5500-watt generator will power needed functions in most homes.

If you are the Do-It-Yourself type or want to get an idea of watt usage, to determine general pricing, without using an amp meter, here is one way to figure your electricity use. Most appliances will display watts or volts and amps. If your appliance only displays volts and amps, simply multiply the volts by the amps and you get watts. For example, 120 volts x 6 amps = 720 watts. Some devices will display VA, which simply means the total of volts x amps. For example, 75 VA = 75 watts.

Instead of upsizing your generator to handle an electric heater, you might consider using infrared space heaters on some emergency outlets for heating. A wood or gas fireplace with a fan is also a popular heating option. Remember, you only need to run the fan on your furnace to distribute heat around your home when heating with a fireplace. Emergencies are circumstances where a ground source heating/cooling system shines, because it not only provides your heating/cooling needs with little electricity per unit of heat/cold delivered, but it can also provide most of your hot water needs.

Appliance	Rated Running Wattage	Appliance	Rated Running Wattage
Freezer	500	Toaster	1650
Furnace Fan-1/2 HP	875	Vacuum	1100
Lights	40-100	Iron	1200
Refrigerator	500	Microwave	800
Television	500	Radio	300
Bug Light	100	Space heater	1800
Coffee Maker	1500	Washing Machine	1150
Computer	720	Electric Dryer	5400
Dehumidifier	650	Heat Pump	4669
Dishwasher	700	Sump Pump-1/2HP	1050
Electric Blanket	400	Sump Pump-1/3HP	800
Electric Fan	800	Well Pump-1/2HP	1000
Electric Frying Pan	1300	Well Pump-1/3HP	750
Electric Range	2100	Window Air Conditioner	1200
Electric Oven	3410	Central Air Cond. 10,000 BTU	1500
Electric Water Heater	4000	Central Air Cond. 20,000 BTU	2500
Garage Door Opener-1/3 HP	750	Central Air Cond. 24,000 BTU	3800
Garage Door Opener-1/4 HP	300	Central Air Cond. 32,000 BTU	5000
Gas Dryer	1800	Central Air Cond. 40,000 BTU	6000
Hair Dryer	1200	Figure 2.44 Appliance wattage	

To help you calculate what electrical devices you would like to include in a back-up/ emergency system, we have listed in Figure 2.44 the watts most devices use. However, other than using an amp meter, nothing is more accurate than looking at the label on your device or appliance to determine watts.

The Chart indicates running watts only; appliances with motors can have a surge when starting that can be several times the running watts. These surges are microseconds to maybe a couple of seconds. To compensate for surge capacity, the rule of thumb applies here too, figure 1 ½ times (150%) the running watts to size your generator.

119

It is important to note, (In Figure 2.45), those appliances which have a rating of over 3000 watts. If you have an electric hot water heater, electric dryer, electric furnace or heat pump, or you must have central air conditioning in use during a power outage, then you will need to upsize your generator to handle these heavy electrical loads. This upsizing can be costly, but can be avoided.

While the most accurate thing to do is to have a professional design your backup/emergency generator system, you can still get an idea of how many watts the appliances that you want to have available in a power outage consume.

GENERATOR SYSTEM COSTS

Most all generators have a label that displays their electrical generating capacity in watts. Figure 2.45 name plate shows that it is a 5000 watt generator. The amount of watts your generator can produce will determine what you can power in your home. Generators of higher output capacity have higher price tags.

A simple manual/pull-start generator of 5500-6000 watts can cost as little as $700.00, but don't forget to add the price of the materials and electricians labor. This could add up to about $1700, including the generator. For this price, you could have an electrician wire a manual transfer switch to your emergency

Figure 2.45 The 5000 watt capacity of this generator is easily seen on

panel. An extension cord from an inexpensive generator could be connected to your transfer switch to provide emergency power to your home. (A materials and installation instructions kit is a great option)

However, an automatic system can start at approximately $4,500.00. This type of system is worth the price, especially if you are out of town frequently. You will never have to worry about losing the contents of your freezer or being vulnerable because your security system is off. We will discuss more about the advantages of both the manual and automatic generators a little later.

It is better for you to have some knowledge of energy systems and appliances so you can best choose what works for you. Knowledge truly is power when deciding on any kind of electricity source for your home. We find that some contractors will sell you a larger system than you may need in order to make more money.

Case in point; we looked at an emergency generator system that a widow had installed in her home. The generator was more than twice the size she needed; therefore, she spent twice the amount she should have. A simple preliminary design would have saved her more than $3,000.00.

Having your system designed and sized independent from the installer is wise and can save you significantly. We know that there are both honest and dishonest contractors. They are in business to make money, and some will sell you a larger system than you need.

MANUAL GENERATOR SYSTEM

We want to show you how to use a generator that you may already have. You may not have thought about how you could use it, in an emergency, to power needed electrical devices in your home. Do you have a generator sitting in the garage collecting dust or one in your motor home that you only occasionally use? You can give your generator an extra job to exercise it and keep it from clogging up, while providing emergency backup power for your home.

Take a look at the drawing in Figure 2.46 to see how you can use your generator to power an emergency panel in your home. This is a basic diagram, but it gives you a good understanding of what a system should include. All of the components are necessary to safely and easily switch from the electrical grid to a backup generator.

First of all, you should know that kits are available for both RV generators and portable gen-

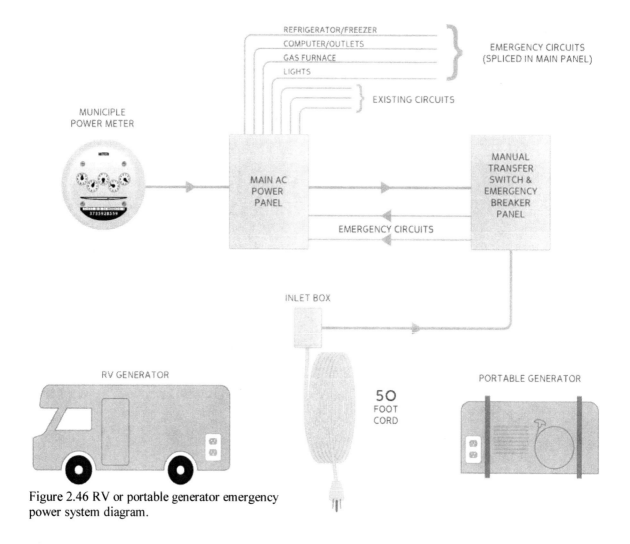

Figure 2.46 RV or portable generator emergency power system diagram.

erators that are sized for different watt capacities. You can go to: MyPowerfulHome.com to find our patented kits. After much research and practical experience we decided to develop kits that were easy enough for the average person to install (depending on state regulations), with the help of quality instructional videos. At the very least, our kits come with such clear instructions in diagram and video, that you will know when an electrician installs it correctly.

A good kit should include:
- Detailed written and video installation instructions and diagrams
- A combination, manual transfer switch/emergency panel

- Breakers
- A properly sized extension cord with a cord cap to match your generator
- An inlet box
- Miscellaneous material
- The estimated cost of the complete installation
- Generator maintenance tips
- System operation instructions

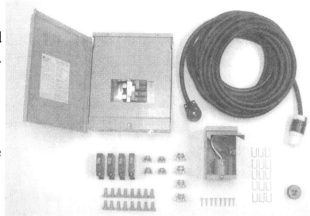

Figure 2.47 Small portable generator kit contents

Just as we discussed in Generator Sizing, you should determine which electrical appliances you want powered in an emergency and route these circuits from your existing panel to your new emergency panel. Make sure that your wattage can be handled by the generator. Generators usually have the watt output written on them, such as 5000 watts.

Next, install your manual transfer switch/emergency panel, place your inlet box as close to your RV or home generator location as possible, and hook up your extension cord with matching plug. Always use a generator outside to avoid carbon monoxide poisoning.

Unless there is a power outage, your manual transfer switch will always be in the *normal position*, supplying power to the circuits that you have moved to the emergency panel from your existing home panel. You will want power to these circuits when the utility company is supplying power and when there is a power outage. In the *emergency position,* your transfer switch will route power from your generator to your new emergency panel, disconnecting it from the main house panel/utility power.

There are two important reasons why you should have an emergency transfer switch:

- First, during most power outages, the power company's employees are generally repairing live power lines. Your little generator can back feed voltage to the transformer and out onto the power grid. By back feeding to a transformer, the voltage is increased instead of decreased (as for home use). Now, you have generated 7200 volts or more to the transmission lines, that is unknown to employees working on the outage issue. This can be deadly.

- Secondly, different AC generators will not synchronize electricity without expensive equipment. If you try to combine your generator's electricity with that of the utility company without synchronizing equipment, something will be damaged. Transfer switches are relatively inexpensive and absolutely necessary because they will not allow both power sources to combine.

AUTOMATIC STANDBY GENERATOR SYSTEM

We have talked about using your RV or home generator for emergency use. Both of these generators (usually AC) require attention in the event you lose power. In addition, like any motor, you should exercise them occasionally to avoid future problems. Neither of these issues are a concern when you have an automatic standby generator system.

Automatic generator systems are more expensive, but much more convenient, especially if you are away from home during a power outage. They are also handy when you don't want to go out in extreme weather conditions to start your generator and pull your manual transfer switch.

Figure 2.48 Automatic standby generator.

An automatic standby generator system exercises itself weekly, starts when the power goes out, and transfers back to the utility power when the power failure is corrected. You can be asleep in your home or out of the country and not notice the change. The smooth transition from utility power to generator power will keep your alarm system and freezer running. Some people have had a quick freeze during a power outage and ended up with significant water damage. This can be avoided with an automatic generator system.

Notice in figure 2.49 that we are showing a dual fuel application for an auto start emergency generator. The power is routed just like the manual transfer switch, except the automatic transfer switch changes it over for you.

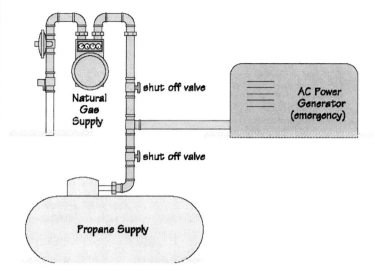

Figure 2.49 Automatic standby power system .

Most home auto-start generators, for emergency use, are natural gas. Many generators have the ability to run on propane by moving a switch by the carburetor or by changing the fuel orifice. Many generators come with orifice kits that can be installed for either gas type.

GENERATOR FUEL TYPES

Some emergency generators are multi-fuel and can easily be switched from natural gas to propane if desired. In Figure 2.50 we show an example of an emergency generator system we provided for a customer that desired two types of fuel sources.

This customer was concerned about an earthquake or some other event disrupting natural gas supply and he wanted the capability of switching to propane. He has several propane tanks that can fuel his generator once one is attached and the valves are turned. He can power his

Figure 2.50 Dual fuel piping.

home's emergency electrical system for days in any event. This gave him the peace of mind that he wanted.

Each generator fuel type has advantages and disadvantages. Let's look at four fuel types: Natural gas, Propane, Diesel, and Gasoline.

Natural gas: This is the most popular fuel generator for emergency systems. It requires no storage tank and runs very clean.

Propane fuel: A propane generator is great for remote locations with no natural gas available. It requires fuel storage but propane can be stored for years. This type of fuel runs clean.

Diesel fuel: High torque diesel engines usually last far longer than other fuel engines. This type of generator can be a good option if you are storing diesel for a diesel vehicle anyway. However, they are more expensive than most other generator types.

Figure 2.51 Manual fuel generator.

Gasoline fuel: This can be a good option for a fuel generator because gasoline is readily available in most places. You could even siphon fuel from your vehicles (Even your neighbor's vehicles can be a resource, ask first.) and you are still in business. This type of generator requires a little more maintenance and fuel storage. Gasoline stores only marginally well, even if you add a fuel conditioner.

AC & DC GENERATORS

AC Generators: AC or alternating current is what you use in your home. Most generators used for emergency power are AC. When you have one of these installed in your home, either manual or automatic, there is no need for an inverter because it produces the same AC current that you are already using. An AC generator generally has the same voltage as your home

Figure 2.52 Plug-in directly to an AC generator.

power (120/240 volts) and can be sized to power the whole house or just a few selected device/appliance circuits as discussed earlier.

These generators can also power your home, for a short period, to assist renewable electricity generators, like solar, wind, or micro-hydro, when there is a need for more electricity than is being produced.

In a net metering scenario with renewable systems, an AC fuel generator can power your home in a power outage. This is because any electricity generator, renewable or fuel, must not supply power to the utility grid in a power outage. A standby or emergency generator system with a transfer switch is used to switch off utility power supply and enable generator supply only.

DC Generators: DC generators are best used in off the grid applications for two main purposes – *battery charging* and *direct power usage*.

Battery charging - Charging your batteries and keeping them on the higher end of a charge extends their life. Since all batteries are DC, a DC generator's electricity generally passes through a charge controller before charging a battery bank. A quality DC generator will modulate its output to charging batteries, depending on the depth of discharge, the type and size of the battery bank, and the demand of the whole DC system.

Direct power usage - When an off the grid home is designed properly, it will have DC lighting and some DC appliances that run directly off the DC panel supplied by the battery bank. This is an energy-saving approach since in the process of inverting DC to AC, a percentage of power is lost (called the "power factor"). A DC generator serves as backup for a battery bank that feeds this type of appliance circuitry.

Your battery bank should be sized to keep the generator run time as low as possible (1 to 2 hours daily), conserving fuel and generator life. Hopefully, your renewable electricity generators are sized to carry most of the power requirements. Some high quality DC generators will run 20 to 40 years when designed properly into a system.

Important Points:

- **Determine which appliances and lighting circuits you want to power and size your generator accordingly.**

- **If you already have a home generator or RV generator, use these for emergency power generation.**

- **An automatic generator system gives you peace of mind when there are power outages and when you are away.**

TYING IT ALL TOGETHER (electrical power)

This is another bird's-eye view with system functions. The following drawing represents all of the electrical generators we have discussed, tied together. While we know that this drawing is over-kill and actually combines on and off-grid systems, the purpose is to show you the integration of all possible components. You may have some combination of these in your powerful home system.

In the following illustration (Figure 2.53), you will see AC and DC electrical panels. The DC panels apply more appropriately to an off-grid system. Essentially, DC current is easier to manipulate for home energy applications. For example, while your renewable electricity generators (solar, wind, micro-hydro, etc) are charging your DC batteries, these generators could also be supplying electricity to DC lighting and DC appliance circuits. This is an efficient use of DC because every time you invert this electricity from DC to AC, you are subject to the *power factor*.

Turn to the next page for a large two page lay out of this diagram and description.

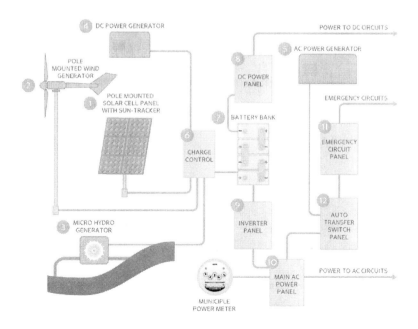

Figure 2.53

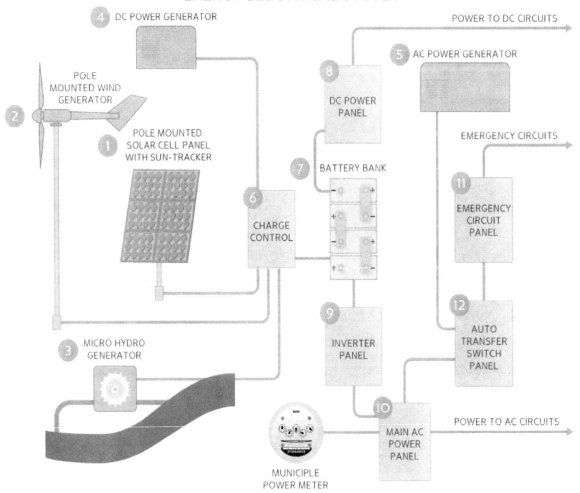

WHOLE HOUSE HYBRID RENEWABLE
ENERGY ELECTRICAL SYSTEM

For a brief summary and a quick review, we will describe component functions as we tie it all together for you.

1. A sun tracking solar array in this system is set for maximum power by moving with the sun and maintaining a 90-degree angle all day long, increasing efficiency by up to 50%.

2. The wind generator is installed above all trees and obstructions and is positioned and sized to capture all of the available wind.

3. A stream has enough fall or drop in elevation to allow a micro-hydro generator to be installed. This generator provides 24-hour electricity, which allows a savings on other renewable generators. The micro-hydro generator also benefits the battery bank by keeping batteries on the upper end of their charge.

4. The purpose of the DC fuel generator is to ensure that the batteries are kept above 80% of full charge, there is an ability to meet high electrical demand, and there is power when other parts of the system are down for maintenance.

5. This backup emergency AC generator provides electricity during a power outage and will keep all of your necessary appliances and lighting circuits operating.

6. The combination charge controller/combiner panel receives all of the electrical wires from your DC generators (solar, wind, micro-hydro, and fuel). This power is then regulated to charge your battery bank.

7. The battery bank is sized to serve as a buffer and store enough power to start motors and handle those occasional high electricity use times. This is important because you don't want to size all of your generators according to the greatest power demand, as this would be wasteful and costly. These batteries should be maintained and charged at all times.

8. The DC panel supplies some lighting and appliances that can use DC voltage. This is efficient because when electricity is inverted from DC to AC, you lose some electricity in the process (power factor). When you are off the grid, use the DC voltage as much as you can because it is originally generated this way from most renewable electricity generators.

9. The inverter changes the electricity from, DC to 120/240 volts AC to run standard appliances and loads designed for this common voltage. Some electricity loss happens during this change in power (power factor); however, it is a necessary evil. Purchase inverters with at least a 94% power factor, which translates into a 6 percent loss of power during the change in voltage.

10. The standard AC electrical panel distributes power to various circuits and AC appliances.

11. The AC emergency panel serves needed appliances, lighting, and outlets. In the event of a power outage; you won't lose frozen food, heating, lights, etc.

12. The auto transfer switch automatically switches from utility power to generator power, even when you are not at home. Your home security system will always be on, even in blackout conditions.

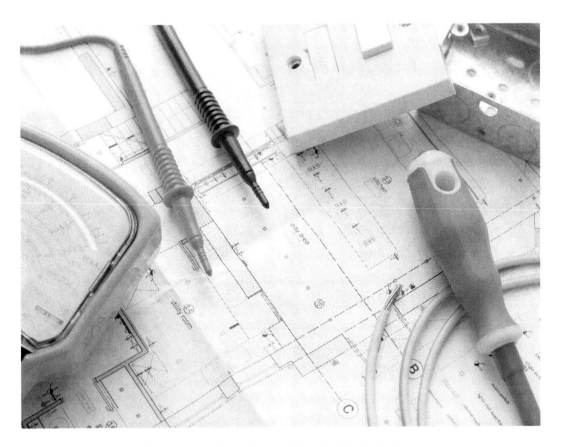

THE RIGHT DESIGN
SAVES YOU MONEY
Renewable Energy Professional Design

THE RIGHT DESIGN SAVES YOU MONEY

Your Questions:

- **Will a professional design safe me money on my renewable energy system?**
- **What should I look for in a professional design?**

Building a home is an expensive and involved undertaking for most people. Generally, people who want to build, realize the importance of having well designed plans. Building plans will help all the trades to coordinate better and will also ensure that building codes are met. Having a professional design for your renewable energy system is no less important.

Figure 2.54 Money saved by professional renewable energy design.

We can hardly count the times that a homeowner has rushed to buy a solar panel kit or some other renewable energy equipment that is too small, too large or just won't work the way they thought it would. Sometimes, people end up spending thousands of dollars to have a professional tradesman correct the deficiencies in their purchases. Here are a couple of true "sad tale" examples:

- We were called in to fix a very nice ground source heat pump system that a doctor had purchased. We were the fourth contractor called in an attempt to get his system to function properly. We were surprised to find that after all the prior work, the installation was not close to working properly. After tearing out an attempted repair, we custom built a control panel, re-piped the flow of transfer fluid and the system worked great. The doctor had to spend an additional $9500.00 to get it right. He would have saved thousands with an initial proper design.

- A college professor purchased a very large emergency generator and called us for the installation. We routed all the lighting, refrigerators, freezers, and outlets that he desired to a new emergency panel, and then installed the auto transfer switch and connected his new generator. While we were showing him how it all worked, we measured the total wattage of the home as opposed to the total wattage of the emergency panel. We discovered that he had overestimated his power usage. A generator system half the size would have powered his whole home not only his emergency panel. What this all boils down to is a more expensive, oversized generator system that could have been avoided had the system been properly designed.

The do-it-yourself (DIY) market is large in renewable energy. All we can say is, "Buyer beware!" We purchased a 'how to' build your own solar panel design book from the internet. After reading it, we wondered how anyone could get a proper and efficient panel built using those instructions. We also wondered how many people bought the same book that we did. If you are a DIY person, get a professional design first and then do it yourself. What may 'look good' on a website or in a catalogue may not be right for your application. Here is one "Buyer beware" example:

- Looking in a catalogue of fuel generators and solar panels, we found some kits that anyone could purchase. One solar system kit boasted 3600 running watts for about $9000.00. It sounded like you would get the capability of producing 3600 watts of electricity from the solar panels. At an average industry price of $5.00 per watt for material alone, that would be a good deal. However, when we read the small print, we realized that it would produce 3600 watts only if the batteries are fully charged and only for a short period. The solar panels in this kit would only produce 420 watts at full sun (90° angle), for 8.6 hours. That would be about $2100.00 worth of solar panels at current average solar panel prices. Even with the price of the battery bank, this was very overpriced.

A PROFESSIONAL DESIGN

When you are looking for a company or a professional that can custom design a home energy efficiency plan and a renewable energy system for your home, look for the following credentials and expertise:

- LEED certification (Leadership in Energy and Environmental Design - efficient building envelope & structure)
- NAHB Green Home Builder certification (National Association of Home Builders - efficient home envelope & structure)
- RESNET certification (Residential Energy Services Network - home energy rating)
- IGSHPA certification (International Ground Source Heat Pump Association - ground source heating & cooling)
- HVAC Journeyman Licensing (Heating Ventilating and Air Conditioning - heating & cooling)
- NABCEP certification (North American Board of Certified Energy Practitioners - solar/photovoltaics)
- Electrical Journeyman or Master Licensing (Licensed electrical professional)
- Many years experience in construction design and management

DESIGN FEATURES

A good custom design will include the following features:
- A home evaluation/energy survey
  - o Includes detailed questions about your home envelope energy consumption and property
  - o Should meet RSNET/HERS standards (Home Energy Rating System)
  - o Can enhance appraisal value with a good HERS rating

- Efficiency suggestions that show percentage of energy reduction and cost savings if implemented
  - o Home Envelope improvements
  - o Insulation
  - o Radiant barriers

      o   Sealing air leaks

      o   Efficient Lighting

      o   Windows and doors

- Recommendations for renewable energy generator types that work at your location (solar, wind, micro-hydro, emergency generator, batteries, ground source heating and cooling)
- Design diagrams and easy to read installation instructions
- A list of system components and approximate costs
- A list of qualified/certified installers in your area
- The approximate cost of an installation
- Tax incentives and rebates available for your location (Federal, state, municipal, and utility company)

The bottom line is, "Knowledge is power". A designed system will give you the knowledge to install a properly sized renewable energy system, and this will save you money.

Important Points:

- **Know a company's credentials and qualifications before using their services.**

- **Save money by having a custom designed renewable energy system.**

- **A good professional design should be thorough and detailed.**

My Powerful Home

SYSTEM COSTS & TAX INCENTIVES
Making Renewable Energy Affordable

SYSTEM COSTS & TAX INCENTIVES

Your Questions:

- **Are there tax incentives for renewable and energy efficient systems & appliances?**

- **Can you show me the cost savings?**

- **Where can I find these incentives?**

Federal, state, and municipal governments, along with many utility companies, are offering incentives to homeowners who install new or upgraded material and equipment that meet efficiency requirements. There has never been a time like now, when so much support is available for homeowners to not only save money using efficient systems, but to be rewarded for installing them.

Figure 2.55 Tax rebates and incentives.

We want everyone to be aware of these incentives and understand how to benefit from them. As you are considering what to install, be sure to calculate all the incentives when totaling the purchase price. As you well know, the price you pay for an appliance, for example, has two parts – what you pay for it and the cost of using it. The current tax incentives and rebates address both of these costs. With the final tally, you might be surprised at what you can afford.

To find tax incentives and local rebates, we suggest that you research the three major government linked sites; EnergyStar.org (For federal tax incentives - renewable energy and home efficiency.), www1.eere.energy.gov/financing/consumers.html (For finding financing and links to several federal sites) and DsireUSA.org (For federal, state and local municipal and utility tax incentives. A great site that is comprehensive for renewable energy and home efficiency tax incentives research.)

The following chart illustrates incentives from one U.S. state chosen at random. This example is representative of the majority of states.

Qualifying Energy Star Appliance	Price range	Example of local utilities rebate
Refrigerator	$400-$3,000	Ames Electric, Iowa $25-$100
Freezer	Upright $450-$1500 Chest $250- $800	Ames Electric, Iowa $50
Dishwasher	$300- $1700	Ames Electric, Iowa $50
Clothes Washer	$450-$1700	Ames Electric, Iowa $100
Clothes Dryer	$480-$1100	No rebate offered, but many other gas utilities do.

Figure 2.56 Energy Star appliances benefits

More and more local utility companies are offering rebates when you buy Energy Star or high efficiency appliances. This will lower the initial price tag of the appliance, but there is an additional cost savings as well. For example, over the life of your new Energy Star qualified washer, you'll save enough money in operating costs to pay for the matching dryer. With your water savings, you could fill three backyard swimming pools. A new Energy Star qualified front-loading washer will use up to 25 gallons of water less, per load, than an older top-loading washer.

It is possible, depending on where your home is, that you could get a Federal and/or State tax credit and a rebate, coupon, or energy credit from your local utility company. Some states will also allow renewable energy equipment property tax exemptions.

The following chart is designed to help you understand that renewable energy can be more affordable by utilizing federal and state tax credits as well as local municipal and/or utility company rebates.

Renewable Energy	Price Range	Fed Tax Credits	Example of State Tax Credits	Example of Local Utilities Rebate	Potential Savings
4 Ton Ground source heat pump, open loop	$6,000-$10,000	30% of system cost	Montana up to $1500	Flat Head Electrical MT, $3,000	$6,300-$7,500
4 Ton Ground source heat pump, horizontal closed loop	$12,000-$20,000	30% of system cost	Montana up to $1500	Flat Head Electrical MT, $3,000	$8,100-$10,500
4 Ton Ground source heat pump, vertical closed loop	$16,000-$28,000	30% of system cost	Montana up to $1500	Flat Head Electrical MT, $3,000	$9,300-$12,900
Active solar Hot water heating system	$3,500-$7,000	30% of system cost	Iowa production tax credit, 1.0₵ to 1.5₵ per kWh	Preston Municipal in Iowa $30/ sq ft of collectible area up to $3,500	$1,650- $5,600
Solar electric (PV system) 2000 watt	$14,000-$20,000	30% of system cost	Iowa production tax credit, 1.0₵ to 1.5₵ per kWh	Preston Municipal in Iowa, $2-$3 per projected kWh up to $10,000	$5,200- $16,000
Wind Generator system 2000 watt	$5,000-$12,000	30% of system cost	Iowa production tax credit, 1.0₵ to 1.5₵ per kWh	Preston Municipal In Iowa, 25% of project up to $10,000	$2,750-$6,600

Figure 2.57 Renewable Energy tax incentives.

When you are researching to better understand what benefits you are eligible for remember that there are two main categories for tax incentives: *Energy Efficiency Tax Incentives* and *Renewable Energy Tax Incentives.*

Energy Efficiency Tax Incentives

This applies to: Biomas Stoves   Heating, Ventilating, Air Conditioning-HVAC

(continued)   Insulation   Roofs (Metal and Asphalt)
              Window and Doors   Water Heaters (non-solar)

Points to remember:

- Tax credit: 30% up to $1500. Any single or combination of the covered items can be used to come up to $1500 total.

  Installation costs are NOT covered for: Windows   Roofs
                                           Doors   Insulation

- Expires December 31, 2010 (As of printing, there is no word on extending or renewing the tax credits for energy efficiency)

- Must be an existing home that you own, your primary residence and be in the United States. It can include a house, houseboat, mobile home, cooperative apartment, condominium, and manufactured home. New construction and rentals do not apply.

Renewable Energy Tax Incentives

This applies to: Geothermal Heat Pumps   Small Wind Turbines
                 Solar Electric System   Fuel Cells
                 Solar Water Heating

Points to remember:

- Tax credit: 30% of Total system cost, generally with no upper limit. Geothermal heat pumps must meet specified minimum EER and COP ratings. Wind turbines must have a nameplate capacity of no more than 100 kW. Fuel cells are capped at $500 per .5 kW of power capacity.

  Installation costs ARE covered for all of the renewable energy items listed above.

- Expires: December 31, 2016

- Existing homes and new construction qualify. Must be your primary residence and be in the United States. It can include a house, houseboat, mobile home, cooperative apartment, condominium, and manufactured home.

There are two sites to research for answers to specific tax questions – EnergyStar.org (Look for the Top Ten Tax Credit FAQs) and the IRS.gov.

Since we are often asked tax questions, we have include the top three FAQs, on the following pages. These come directly from the sites listed above, but do your own research to be sure that you fully understand the question and answer.

1. **Question**

How can I determine if I can collect the tax credit? Does it matter if I am getting a tax refund?

Answer

Whether or not you are getting a refund does not matter. What matters is your "tax liability" - which is the total amount of federal income tax you are responsible for paying.

These energy efficiency tax credits are technically "non-refundable" which means you can't get more money back in tax credits than you pay in federal income taxes (your tax liability). Check your last year's tax return to get a sense of your tax liability (line 61 on the 2008 1040 form, called "total tax"). You can claim all of your tax credits as long as your tax liability, is greater than zero after all tax credits have been applied.

For example, say your Adjusted Gross Income (AGI) is $50,000, your tax liability is $10,000 (before you apply tax credits), and you've had $12,000 withheld from your paychecks. In this scenario you could claim up to $10,000 in tax credits. If you are eligible for the entire $1,500 tax credit, then your tax liability ($10,000) would be reduced to $8,500. Since you already had $12,000 withheld, you will get a tax refund of $3,500 ($12,000 - 8,5000 = $3,500).

If your AGI was $50,000, your tax liability $10,000 (before tax credits were applied), and you had $8,000 withheld from your pay checks, you would still have the ability to claim up to $10,000 in tax credits. If you are eligible for the entire $1,500 tax credit, your tax liability ($10,000)would be reduced to $8,500 and you would owe the government $500 at the end of the year ($8,000 already paid in taxes - $8,500 tax liability = $500 final payment).

Some of the tax credits can be carried forward to future years.

If you don't pay any taxes, then you can't get the credit.

2. **Question**

What is a Manufacture's Certification Statement?

Answer

A Manufacturer's Certification Statement is a signed statement from the manufacturer certifying that the product or component qualifies for the tax credit. Manufacturers should provide these Certifications on their websites. Call the manufacturer, or search their website.

Tax payers must keep a copy of the certification statement for their records, but do not have to submit a copy with their tax return. We have found them easily by 'googling' Manufacturer's Certification Statement and then the name of the company (Owens Corning, for example).-

3. **Question**

Can a tax credit be carried over to future years?

Answer

Only the tax credit for the following products at 30% with no upper limit CAN be carried forward to future years:

Solar Panels   Solar Thermal (water heating)

Answer 3 continued: Geothermal Heat Pumps   Wind Energy Systems

                                     Fuel Cells

If you are unable to claim the entire 30% of your purchase for the above products in one year, you can carry forward the unclaimed portion to future years. The IRS has not issued guidance on how long the tax credit can be carried forward. It is clear that it can be carried forward through 2016, and it appears that it may be able to be carried forward beyond 2016. The Energystar.org site where this came from frequently updates and clarifies points such as this. We encourage you to refer to them.

Important Points:

- **A little research and paperwork adds up to a lot of money.**

- **Use the Dsireusa.org site for your federal and local incentives.**

- **Research the EnergyStar.org site for more tax credit details and product information.**

- **Watch the dates for installation deadlines and for efficiency credits.**

MY POWERFUL HOME

Thank you for following with us as we endeavored to help you understand more clearly the topics of Home Efficiency and Renewable Energy. We are passionate about these topics and enjoy sharing what we know and the resources we have found.

For more information, in the form of videos and articles, go to: www.MyPowerfulHome.com.

You can also follow us on Face Book and Twitter with user names: MyPowerfulHome.

Be one of the first 1000 Fans to sign up with us on Face Book and we will enter you into a drawing for a FREE Emergency Power Kit. This is a $536 value!

GLOSSARY

Absorber: The part of the solar water heater collector that absorbs the sun's energy and changes that energy into heat.

AC (Alternating Current): Current that flows in one direction and then the other, alternately. http://en.wikipedia.org/wiki/Alternating_current

Active Solar Water Heater System: Active solar uses mechanical devices such as pumps and valves to move collected heat from the solar collector panel to storage mediums and/or end use. Thus, solar radiation is used by special equipment to provide our homes with space heating or hot water.

Airfoil: The shape of the blade cross-section, which, for most modern horizontal axis wind turbines, is designed to enhance the lift and improve turbine performance. http://en.wikipedia.org/wiki/Airfoil

Air-Source Heat Pump: A heat pump that uses an air-to-refrigerant heat exchanger to extract and reject heat to the outside air.

Ambient Air: The surrounding air (usually outdoor air or the air in a specific location).

Ambient lighting: Lighting in an area from any source that produces general illumination, as opposed to task lighting.

Amps; Amperes: Amps are a measure of electrical current. In incandescent bulbs, the current is related to voltage and power as follows: Watts (power) = Volts x Amps (current). http://en.wikipedia.org/wiki/Ampere

Anemometer: A device to measure wind speed.
http://www.britannica.com/EBchecked/topic/24295/anemometer

ARI: Air-Conditioning and Refrigeration Institute.

Array: An array includes any number of photovoltaic modules connected together to provide a single electrical output. Arrays are often designed to produce significant amounts of electricity.

Average wind speed: The mean wind speed over a specified period of time.

Balance Point: The temperature above which the heat pump can provide enough heat for the home without the use of supplemental heating.

Ballast: A ballast is a device that serves to control the flow of power to a fluorescent lamp. Advanced electronic ballasts have replaced many magnetic ballasts of the past in new CFL Bulbs and fixtures. Electronic ballasts improve fluorescent energy efficiency even further and eliminate the ''hum'' and visible flickering found in older fluorescent technology. Some are compatible with dimming and daylight controls.

Blades: The aerodynamic surfaces that catch the wind.

Blowers: Blowers are fans used to force air across the heat exchanger. With a ground-source heat pump, the only blower used is to force air through the central heating/cooling system.

Brake: Used in wind turbine systems to stop the rotor from turning.

BTU, British thermal unit: The quantity of heat required to raise the temperature of one pound of water one degree Fahrenheit at a specified temperature.

Building Envelope or Enclosure: The building envelope is the outer shell or the elements of a building such as walls, floors, and ceilings, which enclose conditioned space. The building envelope separates the conditioned space from the unconditioned space or the out-doors.

Cavitation: The formation of bubbles due to partial vacuums in a flowing liquid as a result of separation of fluid particles.

Cellulose Insulation: Cellulose insulation is made from recycled newsprint and other paper sources, and is heavily treated with fire retardant chemicals. http://en.wikipedia.org/wiki/Cellulose_insulation.

CFL, Compact Fluorescent Lamp: The general term applied to fluorescent lamps that are single-ended and that have smaller diameter tubes that are bent to form a compact shape. CFLs are four times more efficient and last up to 10 times longer, 10,000-15,000 hours, than incandescent light bulbs. A 22-watt CFL has about the same light output as a 100-watt incandescent. Compact fluorescent light bulbs use 50-80% less energy than incandescent light bulbs. _http://www.energystar.gov/index.cfm?c=cfls.pr_cfls

Charge Controller: A charge controller is a component of a photovoltaic system that controls the flow of current to and from the battery to protect it from over-charge and over- discharge. The charge controller may also indicate the system operational status.

Check Valve: A valve that allows a fluid to travel in only one direction within a water piping circuit.

Circuit: A path for electricity to flow.

Circulating Pump: The pump(s) that circulate the fluid in the closed-loop system during normal operation.

Clay: Clay is made up of mineral soil particles that are less than 0.002 millimeters in diameter. As a soil textural class, clay is made up of soil material that is 40% sand and less than 40% silt.
http://en.wikipedia.org/wiki/Clay

Closed-Loop System: In a ground-coupled system, a pressurized heat exchanger consisting of the ground heat exchanger, the circulating pump, and the water-source heat pump circulates heat transfer fluid in a closed polyethylene pipe system.

Coil: A coil is a heat exchanger used to transfer energy from one source to another. In ground-source heat pumps, water-to-refrigerant and refrigerant-to-air coils are used.

Collector: A device that collects solar energy.

Compressor: A compressor is the central component of a heat pump system. The compressor increases the pressure of a refrigerant fluid, and simultaneously reduces its volume, while causing the fluid to move through the system. Heat is produced in the process.

Condenser: A heat exchanger in which hot, pressurized (gaseous) refrigerant is condensed by transferring heat to cooler surrounding air, water, or earth.

Conduction: Conduction is the transfer of heat through a material. Heat is transferred directly in and through the substance. Conduction heat loss or gain results from the transfer of heat directly through the materials of the building envelope. If the outside temperature is greater than the inside temperature, there is heat gain from outside the building.

Converter: A device that converts AC to DC electricity.

COP (Coefficient of Performance): Coefficient of Performance is the efficiency ratio of the amount of heating or cooling provided by a heating or cooling unit to the energy consumed by the system. The higher the coefficient of performance, the more efficient the system is. Electrical heating, for example, has a Coefficient of Performance of 1.0. http://homerepair.about.com/od/termsaf/g/COP.htm

Cut-in wind speed: The wind speed at which a wind turbine begins to generate electricity. http://www.energy.eu/dictionary/data/392.html

DC (Direct Current): Electricity which flows in one direction
http://www.school-for-champions.com/science/dc.htm

Desuperheater: A desuperheater is a device for recovering superheat from the compressor discharge gas of a heat_pump or central air conditioner for use in heating or preheating water. This hot water then circulates through a pipe to the home's storage water heater tank. Also known as a heat recover water heater.

DOE: U.S. Department of Energy. http://www.energy.gov/about/index.htm

Electrician Journeyman and Master: Most states require individuals to have 4 to 6 years of on the job electrical experience along with schooling. Most licensed electricians can quickly adapt to PV installations because they generally deal with far more complicated systems. Having said that, PV installations require knowledge about roof and pole systems, battery sizing, solar array sizing, net metering, and other factors unique to the PV industry.

Energy Efficiency: Efficiency is defined as the ability to accomplish a job with a minimum expenditure of time and effort. Energy efficiency is the ability to provide a function or service with the minimum expenditure of energy.

EER (Energy Efficiency Ratio): The EER is a dimensional quantity usually used in specifying cooling performance. The EER is the ratio of cooling provided by the system (in Btu) to the energy consumed by the system (in watt-hours) under designated operating conditions. http://saveenergy.about.com/od/understandingenergy/g/EER_DEF.htm

Energy Star® Home: A home, certified by the U.S. Environmental Protection Agency (EPA), that is at least 30% more energy efficient than the minimum national standard for home energy efficiency as specified by the 1992 MEC, or as defined for specific states or regions. ENERGY STAR is a registered trademark of the EPA.

ERV (Energy Recovery Ventilator): HRVs and ERVS are similar devices that both supply air to the home and exhaust stale air while recovering energy from the exhaust air in the process. The primary difference between the two is that an HRV transfers heat while an ERV transfers both heat and moisture. ERVs are used for warmer, more humid climates with long cooling seasons.

http://www.hvi.org/assets/pdfs/HRV.ERVBrochJune2008.pdf___http://www.iaqsource.com/hrvs_ervs.php

Evaporator: A heat exchanger in which cold, low-pressure (liquid) refrigerant is vaporized to absorb heat from the warmer surrounding air, earth, or water.

Fiberglass insulation: Made from sand and recycled glass.

Fluorescent Lamp: A high efficiency lamp utilizing an electric discharge through inert gas and low-pressure mercury vapor to produce ultraviolet (UV) energy (See Compact Fluorescent also)._http://en.wikipedia.org/wiki/Fluorescent_lamp

Furling: The process of forcing, either manually or automatically, the blades of a wind turbine out of the direction of the wind in order to stop the blades from turning in extreme winds.

Geothermal Heating and Cooling: This is an interchangeable term for Ground Source Heating and Cooling. Grounds source heat pumps take advantage of the earth's heat in the top 300' of the crust. The majority of this geothermal energy comes from the sun.

Grid: The utility distribution system or network that connects electricity generators to electricity users.

Ground-Coupled Heat Pump: A ground-coupled heat pump is one that uses the earth itself as a heat source and heat sink. It is coupled to the ground by means of a closed-loop heat exchanger (ground coil) installed horizontally or vertically underground.

HAWT: Horizontal axis wind turbine.

Heat Exchanger: A heat exchanger is a device built for efficient heat transfer from one medium to another, whether the media are separated by a solid wall so that they never mix, or the media are in direct contact. They are widely used in space heating, refrigeration, air conditioning, power plants, chemical plants, petrochemical plants, petroleum refineries, and natural gas processing. One common example of a heat exchanger is the radiator in a car, in which the heat source, being a hot engine-cooling fluid, water, transfers heat to the air flowing through the radiator.

Heat Pump: A heat pump is a mechanical device used for heating and cooling, which operates by pumping heat from a cooler to a warmer location. Heat pumps can draw heat from a number of sources (e.g., air, water, or earth) and are most often either air-source or water-source.

Heat Sink: The medium (e.g., air, water, earth, etc.) that receives heat from a heat pump.

Heat Source: The medium (e.g., air, water, earth, etc.) from which heat is extracted by a heat pump.

HERS, Home Energy Rating System: A standardized system for rating the energy efficiency of residential buildings

HERS Index: A numerical integer value produced by a Home Energy Rating that represents the relative energy use of a Rated Home as compared to the energy use of the HERS Reference Home, and where an Index value of 100 represents the energy use of the HERS Reference Home and Index value of 0 (zero) represents a home net purchased energy.
A standardized system for rating the energy efficiency of residential buildings.

HRV, Heat Recovery Ventilation: HRVs and ERVS are similar devices that both supply air to the home and exhaust stale air while recovering energy from the exhaust air in the process. The primary difference between the two is that an HRV transfers heat while an ERV transfers both heat and moisture. HRVs are usually recommended for colder climates with longer heating seasons.
http://www.iaqsource.com/hrvs_ervs.php

HVAC: Heating, ventilating, and air conditioning.

HVAC Journeyman: In most states, an HVAC journeyman is a person who has worked a minimum of 4 years along with schooling in the heating, ventilating, and air conditioning trade. Some HVAC journeymen have been trained in ground source heating and cooling without the IGSHPA certification.

Hydroelectric Energy: Electricity generated from the force of moving or falling water. A form of renewable energy.
http://cyberparent.com/green-building/glossary/index.htm

Hydronic: A heating or cooling distribution system using liquid piped throughout the house to radiators or convectors.

IGSHPA (International Ground Source Heat Pump Association): IGSHPA is a nonprofit, member-driven organization established in 1987 to advance ground source heat pump (GSHP) technology on local, state, national and international levels. Headquartered on the campus of Oklahoma State University in Stillwater, Oklahoma, IGSHPA utilizes state-of-the-art facilities for conducting GSHP system installation training and geothermal research. With its access to the most current advancements in the geothermal industry, IGSHPA is the ideal bridge between the latest technology and the people who benefit from these developments. http://www.igshpa.okstate.edu/

Incandescent Light Bulb: is a source of electric light that works by incandescence. An electric current passes through a thin filament, heating it until it produces light. The enclosing glass bulb prevents the oxygen in air from reaching the hot filament, which otherwise would be destroyed rapidly by oxidation. Incandescent bulbs are also sometimes called, electric lamps. They are the most common light bulbs by far.

Inverter: A power inverter converts DC power or direct current to standard AC power or http://www.wisegeek.com/what-is-an-alternating-current.htmalternating current. http://en.wikipedia.org/wiki/Inverter_(electrical)

Joint, Butt-Fused: A joint in which the prepared ends of the joint components are heated and then placed in contact to form the joint.

Joint, Heat-Fused: A joint made using heat and pressure only.

Joint, Socket-Fused: A joint in which the two pieces to be heat fused are connected using a third fitting or coupling with a female end.

Kilowatt (kW): a measure of power for electrical current (1000 Watts).

Kilowatt Hour (kWh): The standard measure of electrical energy and the typical billing unit used by electrical utilities for electricity use. A 100-watt lamp operated for 10 hours consumes 1000 watt-hours (100 x 10) or one kilowatt-hour. If the utility charges $.10/kWh, then the electricity cost for the 10 hours of operation would be 10 cents (1 x $.10). http://en.wikipedia.org/wiki/Kilowatt_hour

Leeward: away from the direction from which the wind blows.

LED, Light Emitting Diode: An efficient source of electrical lighting, typically lasting 50,000 to 100,000 hours.

Loam: Soil material that is 7% to 27% clay particles, 28% to 50 % silt particles, and less than 52% sand particles. http://en.wikipedia.org/wiki/Loam

Incandescent Light Bulb: An incandescent light bulb is a source of electric light that works by incandescence. An electric current passes through a thin filament, heating it until it produces light. The enclosing glass bulb prevents the oxygen in the air from reaching the hot filament, which otherwise would be destroyed rapidly by oxidation. Incandescent bulbs are also sometimes called electric lamps.

Inverter: A power inverter that converts DC power or direct current to standard AC power or alternating current. http://en.wikipedia.org/wiki/Inverter_(electrical)

Kilowatt (kW): A measure of power for electrical current (1000 Watts)

Kilowatt-hour (kWh): The kilowatt-hour is the standard measure of electrical energy and the typical billing unit used by electrical utilities for electricity use. A 100-watt lamp operated for 10 hours consumes 1000 watt-hours (100 x 10) or one kilowatt-hour. If the utility charges $.10/kWh, then the electricity cost for the 10 hours of operation would be 10 cents (1 x $.10). http://en.wikipedia.org/wiki/Kilowatt_hour

LED, Light Emitting Diode: An efficient source of electrical lighting, typically lasting 50,000 to 100,000 hours.

LEED: The Leadership in Energy and Environmental Design (LEED) Green Building Rating System encourages and accelerates global adoption of sustainable green building and development practices through the creation and implementation of universally understood and accepted tools and performance criteria. LEED is an internationally recognized green building certification system, providing third-party verification that a building or community was designed and built using strategies aimed at improving performance that matter most in a building structure. LEED recognizes energy savings, water efficiency, CO_2 emissions reduction, improved indoor environmental quality, and stewardship of resources and sensitivity to their impacts.

Loam: Soil material that is 7% to 27% clay particles, 28% to 50 % silt particles, and less than 52% sand particles. http://en.wikipedia.org/wiki/Loam

Low-E Glass: Low-E glass has the ability to allow visible light to pass while blocking certain amounts of UV light and IR light. Infrared light is basically heat. The infrared light in sunlight is powerful. When it strikes an object, it heats it up. These objects can be your tile floors, furniture, sidewalks, patio furniture, etc. As these objects cool off, they emit a low powered form of IR light. Low-E glass reflects this form of energy. In the summer, this helps to keep your house cooler, as the heat from objects outside is kept outside. In the winter, all objects in your home are heated (by either the sun or your furnace). This heat is also bounced back into your house by the low-E glass.

Micro-hydro: A term used for hydroelectric power from a smaller type of turbine/generator

Multi-zone Heat Pump: A central, split-system heat pump consisting of an outdoor heat exchanger, compressor unit, and multiple (three to five) indoor air handling units, each of which can be independently controlled and operated for space conditioning

MW (Megawatt): A measure of power (1,000,000 watts)

NABCEP (North American board of certified energy practitioners): PV installer certification is a voluntary certification that provides a set of national standards by which PV installers with skills and experience can distinguish themselves from their competition. Certification provides a measure of protection to the public by giving them a credential for judging the competency of practitioners. It is not intended to prevent qualified individuals from installing PV systems or to replace state licensure requirements.

NAHB Green builders: The NAHB employs a National Green Building Program. The National Association of Home Builders is helping its members move the practice of green building into the mainstream. Energy efficiency, water and resource conservation, sustainable or recycled products, and indoor air quality are increasingly incorporated into the everyday process of home building. The National Green Building Program offers several resources and tools to help builders, remodelers, home building associations, and homeowners learn how to build green, and the benefits of doing so.

Net Metering: When a net metering customer's renewable generator is producing more power than is being consumed, the electric meter runs backward, generating credits. When a net metering customer uses more power than is being produced, the meter runs forward normally. Net metering customers are charged only for the "net" power that they consume from the electricity service provider that has accumulated over a designated period or, if their renewable energy-generating systems make more electricity than is consumed, they may be credited or paid for the excess electricity contributed to the grid over that same period.

Open Loop System: A ground source heat pump system in which a heat exchange/transfer fluid is pumped from the ground from a well and then is discharged into another well or into a reservoir.

Parallel System: A flow condition where two or more fluid paths are possible in the closed-loop pipe circuit

Permeability: Permeability is the quality of the soil that enables water to move downward through the profile. Permeability is measured as the number of inches per hour that water moves downward through the saturated soil.

Pipe Heat Transfer Resistance: The resistance to heat flow resulting from pipe wall thermal properties and dimensions

Potable water: Drinkable water

Pressure: When expressed with reference to pipe, the force per unit area exerted by the medium in the pipe

Pressure Rating: The estimated maximum pressure that the medium in the pipe can exert continuously with a high degree of certainty that failure of the pipe will not occur

Propylene Glycol: Propylene glycol is a nontoxic antifreeze used in solar heating systems and ground source heating and cooling systems. It is mixed with water to prevent the freezing of the transfer fluid.

Pump Kit: A prefabricated pumping and service unit for the closed-loop circuit of a closed-loop/ground-source (cl/gs) heat pump system

PURPA: The Public Utility Regulatory Policies Act (1978) refers to small generator utility connection rules.

R-Value: R-Value is a measure of how well an insulation product resists the flow of heat or cold through it. A higher number indicates that the object or material has a higher insulating capability.

Rated output capacity: The output power of a wind machine operating at the rated wind speed

Rated wind speed: The lowest wind speed at which the rated output power of a wind turbine is produced

Refrigerant: A refrigerant is a fluid of extremely low boiling point used to transfer heat between the heat source and heat sink. It absorbs heat at low temperatures and low pressures and rejects heat at higher temperatures and higher pressures, usually involving changes of state in the fluid (from liquid to vapor and back).

Renewable Energy: Renewable energy, quite simply, is energy derived from natural resources that are naturally replenished. Renewable energy includes the following: Wind power, wave power, tidal power, solar power, hydropower, geothermal power, biomass, and biofuel. http://www.biggreensmile.com/green-glossary/wind-power.aspx

The Residential Energy Services Network (RESNET®): RESNET's mission is to ensure the success of the building energy performance certification industry, set the standards of quality, and increase the opportunity for ownership of high performance buildings. RESNET is a membership 501-C-3 nonprofit organization. RESNET's standards are officially recognized by the U.S. mortgage industry for capitalizing a building's energy performance in the mortgage loan, certification of "White Tags" for private financial investors, and by the federal government for verification of building energy performance for such programs as federal tax incentives, the Environmental Protection Agency's ENERGY STAR program, and the U.S. Department of Energy's Building America Program.

Return (Air): Air returned to the space-conditioning unit from the conditioned space

Rotor: The rotating part of a wind turbine, including either the blades and blade assembly or the rotating portion of a generator

Rotor diameter: The diameter of the circle swept by the rotor

Rotor speed: The revolutions per minute of the wind turbine rotor

RPM, Revolutions per minute: The number of times a shaft completes a full revolution in a minute

Sand: Sand includes coarse-grained soil with a size range specified by the classification system used. Sands are non-plastic and non-cohesive. http://en.wikipedia.org/wiki/Sand

SEER (Seasonal Energy Efficiency Ratio): A measure of seasonal cooling efficiency under a range of weather conditions assumed to be typical of a location, of performance losses typical of a location, and of performance losses due to cycling under part-load operation http://en.wikipedia.org/wiki/Seasonal_energy_efficiency_ratio

Series System: A system in which the circulating fluid from the heat pump(s) has a single flow path through the ground heat exchanger

Sieve: A standard woven-wire square-mesh utensil used to strain or sift a soil sample

Silt: Silt includes fine-grained soil with a size range specified by the classification system used. Silts are non-plastic and non-cohesive. http://en.wikipedia.org/wiki/Silt

Soil/Field Resistance: The resistance to heat flow resulting from soil thermal properties and underground pipe placement

Solar Energy: Radiant energy from the sun

Solar Irradiance: Total solar irradiance describes the radiant energy emitted by the sun over all wavelengths that fall each second on 1 square meter outside the earth's atmosphere – a quantity proportional to the "solar constant" observed earlier in this century. It measures the solar energy flux in watts/square meter.

Solar Panel, Photovoltaics: Solar panels, also known as photovoltaics, are used to convert light from the sun, which is composed of particles of energy called "photons," into electricity that can be used to power electrical loads.

Start-up wind speed: The wind speed at which a wind turbine rotor will begin to spin. See also Cut-in wind speed.

Supplemental Heating: Supplemental heating is a heating system component used when a heat pump is operating below the balance point, during defrost, or as an emergency backup when the main system is inoperable. Usually electric resistance heat is used, but natural gas, LPG, or oil heating systems are also used.

Swept area: The area swept by the turbine rotor

Therm: A quantity of heat equivalent to 100,000 Btu

Thermostat: An instrument that responds to changes in temperature and is used to directly or indirectly control indoor temperature by operating a space conditioning system

Ton of Refrigeration: A measure of cooling delivered by a heat pump (or other air conditioning system) equal to 12,000 Btu per hour

Tracking Array: PV array that follows the path of the sun to maximize the solar radiation incident on the PV surface

Turbulence: The changes in wind speed and direction, frequently caused by obstacles

Valve, Expansion: A device for regulating the flow of liquid refrigerant to the evaporator. Two types of valves are commonly used: an electronic valve that responds to variation in electric resistance reflecting changes in refrigerant temperature, and a thermostatic valve that uses a refrigerant-filled bulb to sense changes in refrigerant temperature.

Valve, Reversing: An electrically operated valve that allows the heat pump to switch from heating to cooling, or vice versa, by changing the refrigerant's direction of flow

Volt: A volt is a measure of "electrical pressure" between two points. The higher the voltage, the more current will be pushed through a resistor connected across the points. The volt specification of an incandescent lamp is the electrical "pressure" required to drive it at its designed point. The "voltage" of a ballast (e.g. 277 V) refers to the line voltage to which it must be connected.

VAWT (Vertical Axis Wind Turbine): A wind generator design where the rotating shaft is perpendicular to the ground and the cups or blade rotates parallel to the ground

Water-Source Heat Pump: A heat pump that uses a water-to-refrigerant heat exchanger to extract heat from the heat source

Water-Source Heat Pump, Closed Loop: Closed-loop systems circulate a heat transfer fluid (such as water or a water-antifreeze mixture) continuously to extract or reject heat from a ground or water heat source or sink.

Water-Source Heat Pump, Open Loop: Open-loop systems pump groundwater or surface water from a well, river, or lake through a water-to-refrigerant heat exchanger and return the water to its source, a drainage basin, pond, or storm sewer.

Watts: Watts are a unit of electrical power. Lamps are rated in watts to indicate the rate at which they consume energy.
 http://www.biggreensmile.com/green-glossary/biofuel.aspx

Wind farm: A wind farm is a group of wind turbines, often owned and maintained by one company. It is also known as a wind power plant.

Wind Generator: A device that captures the force of the wind to provide rotational motion to produce power with an alternator or generator

Wind Turbine: A wind turbine is a machine used to capture the force of the wind. It is called a wind generator when used to produce electricity, a windmill when used to crush grain or pump water.

INDEX

Air Conditioning, 12, 32, 40, 56, 120, 136, 142, 152-3, 159

Air leaks, 2, 3, 4, 6-8, 13-14, 137 *See* Chimney effect

Air source heating and cooling, 34, 41, 45-46, 53

Alternating Current (AC), 56, 62-66, 105, 113, 117, 124, 126-132, 148, 150, 154-155

Amps,/Amperage, 65, 66-67, 73-76, 148

Appliances, 9, 27-30, 59, 62, 64-65, 73, 83, 88, 117-123, 127, 129-132, 140-141

Attic/Loft, 2, 4-9

Basement, 2, 4-6, 16, 40

Batteries, 64, 70-79

   alkaline batteries, 71

   battery bank size, 70, 73

   battery charging, 70-71, 75, 127

   battery life, 70, 74

   charge controllers, 72-76, 127, 131,149

   lead acid batteries, 71-73,79

   types, 70-72

   Vs. net metering, 72

Chimney effect, 2, 4, 7-8, 11

Cooling, ix, 14, 31-49, 53, 56, 118, 136-137, 149-153, 157-160

Department of Energy (DOE), ix, 28-29, 53, 101, 151

Desuperheaters, 38, 41-43, 47-48, 52-53, 56, 151

Direct Current (DC), 59-66, 70, 73, 75, 82, 84, 105, 113, 116-117, 126-129, 131, 151, 154-155

Drafts. *See* Air leaks

Dryer. *See* Appliances

Electricity, *See* Alternating Current (AC), *See* Amps, *See* Direct Current (DC), *See* Emergency back-up power generators, *See* Inverters, *See* Micro-hydro power, *See* Net metering, *See* Wind power, *See* Solar power, *See* Storing electricity, See Ttransformers, *See* Volts, *See* Watts.

Emergency back-up power generators, 115-116

   AC generators, 66, 124,126, 131

Emergency back-up power generators - continued

   automatic standby generator systems, 117,124

   DC generators, 116-117,126-127,131

   manual transfer switch, 120-127, 132, 135

   fuel types, 117, 125-126

   generator sizing, 117, 123

   genrator system costs, 117,120

   portable generators, 122-123

   RV generators, 121, 128

Electric Commutated Motor (ECM), 41-43, 56

Energy Efficiency, viii, ix, 9, 19, 45, 136, 142-144, 149, 151, 153, 156, 158

Energy Star, ix, 7, 9-10, 19-20, 28-32, 46, 83, 141, 151, 158 *See* Appliances, *See* Lighting

Expanded Polystyrene (EPS) foam, 12, 14-15

Geothermal heating and cooling. *See* Ground source heating and cooling

Ground heat exchangers, 34, 43, 150, 158

   open loop systems, 35-37, 156, 160

   closed loop system, 35-37, 156, 160

Ground Source Heat Pumps (GSHPs), 13, 33, 37, 41-42, 47-48, 54-58, 134-136, 153-154, *See* Desuperheaters

   combination (water to water and water to air), 40-41

   split systems, 31, 41

   water to air, 37

   water to water, 39

Ground source heating and cooling, vii, 1, 32-34, 37, 43-44, 47, 53, 152-153, *See* Air source heating and cooling, *See* Ground heat exchanger, *See* Ground Sourse Heat Pumps (GSHPs)

Heat Pump Water Heaters, 47, 53, 55, 167

Heating, ix, 13-14, 21, 32-56, 118, 132, 136-137, 142-143, 145, 148-157, 159-160, 167, 169

Heating Ventilation & Air Conditioning (HVAC), *See* Hybrid heating and cooling, *See* Ground source heating and cooling

Home envelope, ix, 2, 4-5, 8, 12, 83, 96, 136

Hot water, ix, 32-34, 38-43, 47-50, 52, 54, 56-58, 118, 120, 148, 151

Hot water production, ix, 31, 43, 47, 55, *See* Desuperheaters, *See* Heat Pump Water Heaters (HPWH), *See* Instantaneous water heaters, *See* Solar water heating

Hot water storage, 48-49, 54-55

Hybrid heating and cooling systems, 32, 45-46

Insulated Concrete Forms (ICF), 12-16

Insulation, 2, 45

   attic/loft, 6

   basement/cellar, 5

   recessed light fixtures, 8

   reflective barriers, 7

   walls, 6

   windows and doors, 9

Inverters, 62-65,87-88,113,126,131,154-155

   micro inverters, 65

   whole system inverters, 65

Lighting, ix, 9, 18-25, 71, 83, 88, 117, 127, 129, 131-132, 135, 137, 148, 154-155

   comapact flourescent lighting (CFL), 18-23, 149, 168

   heat vs. light, 21

   incandescent lighting, 18-23, 148-149, 154-155, 160

   LED lights, 18, 20-25, 76, 154

   true bulb cost comparison, 23

   Tubular Daylighting Devices (TDDs), 24

Manufacturer's Certification Statement, 145

Micro-hydro power, 59, 65, 103-105, 109-110, 113, 131

   basics, 104

   calculating power, 108

   grid-tied systems, 112

   head pressure, 104-108, 113, 169

   measuring flow, 104, 108

   off-grid systems, 112

   pipeline, 110

   pressure gauge, 106-107, 111

   settling tank, 110

Micro-hydro power - continued
   shut off valve, 111
   tailrace, 110, 112
   water intake, 112
Money saving, 134, 140, *See* Tax incentives
Net metering, 59, 63-64, 69-71, 77-78, 98, 127, 151, 156
   agreements, 63-64, 77
   benefits, 77-78
   interconnection standards, 77-79
   problems, 78
Photovoltaics, 81-84, 136, 148-149, 159
   amorphous, 84-85
   basics, 83
   Monocrystalline, 84-85
   polycrystalline, 84-85
Professional design, 59, 113, 120, 134-136,
   certifications, 136
Recessed light fixtures, 8
Reflective barriers, 7
Refrigerator. *See* Appliances
Solar power, 59, 81-90, 158, *See* Photovoltaics
   efficiency optimization, 59, 81, 86
   off the grid with solar power, 88
   upcoming technology, 90
Solar water heating, 48, 55, 82, 143, *See* Hot water storage
   active solar water heaters, 51
   Single tank system, 49-50
   two/dual tank system, 50
Storing electricity, 59, 69-71, 76-77, 83, 88, 96, 98, *See* Batteries, *See* Net metering
Tax incentives, viii, 5, 9, 11, 45, 52, 137, 139-143, 158
   energy efficiency incentives, 9, 142
   Renewable energy, 143
   tax credits, 5, 52-53, 83, 137, 140-142, 167

Utilities rebates, 125

The Structural Insulated Panel System (SIPS), 12, 15-16

Transformers, 62-64, 66, 123

Volts, 59-65, 73, 84, 118, 124, 131, 148, 160

Walls

    existing home, 6

    Expanded Polystyrene (EPS) foam, 12, 14-15

    Insulated Concrete Forms (ICFs), 12

    new construction, 6, 12-16

    The Structural Insulated Panel System (SIPS), 15-16

Washer. *See* appliances

Watts, 64-65, 71, 92-99, 101-102, 143, 148-152, 157-160

Wind power, 59, 91-92, 110, 158, 160, *See* wind turbines

    air density, 92-93

    Anemometer, 93, 95, 148

    blade area, 92-93

    obstructions, 93-94

    wind measurement, 95

Wind turbine, 64-65, 71, 92-99, 101-102, 143, 148-149, 151-152, 157-160, *See* Wind Power

    height, 101

    horizontal axis turbines, 98, 148, 152

    tower types, 99-100

    turbine sizing, 96-98

    Vertical axis wind turbines, 98-99, 160

Windows and doors, 8-10, 137

    boosting efficiency, 8

    replacing, 9

<u>RESOURSES</u>

Appliances
U.S. Department of Energy; Appliance calculator
http://www.energystar.gov/index.cfm?fuseaction=refrig.calculator
U.S. Department of Energy; Rebates
http://www.energystar.gov/index.cfm?fuseaction=rebate.rebate_locator
U.S. Department of Energy; Energy Star Appliances
http://www.energystar.gov/index.cfm?c=appliances.pr_appliances

Generators
Cutler-Hammer: Sizing a Generator for your needs
https://www.ch.cutler-hammer.com/generatorCalc/wattshow.jsp

Heating & Cooling
Geocomfort
http://www.geocomfort.com/geothermal-technology
http://www.youtube.com/watch?v=OzRLPf4tBN0
International Ground Source Heat Pump Association
http://www.igshpa.okstate.edu/geothermal/faq.htm
Alternative Energy; Geothermal Heat pumps
http://alternativeenergy.com/profiles/blog/show?id=1066929%3ABlogPost%3A46637
Canadian Geothermal
http://www.canadiangeothermal.com/
The Geothermal Company of Illinois; Geothermal loop designs
http://www.geoillinois.com/geo.cfm?geopage=4
Indiana Chapter of The Nature Conservancy
http://www.nature.org/wherewework/northamerica/states/indiana/files/geothermal.pdf
Choice Mechanical; Radiant floor heating
http://choicemechanicalinc.com/residential.html
U.S. Department of Energy; Selecting a heat pump
http://www.energysavers.gov/your_home/space_heating_cooling/index.cfm/mytopic=12670
http://www.energysavers.gov/your_home/space_heating_cooling/index.cfm/mytopic=12700
Solar hot water
http://www.energystar.gov/index.cfm?c=solar_wheat.pr_how_it_works
Heat Pump water heaters
http://www.energysavers.gov/your_home/water_heating/index.cfm/mytopic=12840
Air-Conditioning, Heating and Refrigeration Institute; Choosing a hot water heater
http://www.ahrinet.org/ARI/util/showdoc.aspx?doc=1516

Home Envelope

Energy Use Pie Chart
http://www1.eere.energy.gov/consumer/tips/home_energy.html
U.S. Department of Energy; Common Air Leaks
http://www.energystar.gov/index.cfm?c=home_sealing.hm_improvement_sealing

Insulation

U.S Department of Energy
http://www.energystar.gov/index.cfm?c=tax_credits.tx_index
http://www.ornl.gov/sci/roofs+walls/insulation/ins_16.html
Owens Corning; Insulation Installation Instruction Video
http://www.youtube.com/user/owenscorning
Photo of Various Insulation types
http://healthyhomesuk.com/Solar_Energy/Sustainable_Energy/sustainable_energy.html
Innovative Insulation Inc.; Attic radiant barrier
http://www.radiantbarrier.com/energy-savings.htm
Tenmat Inc.; Insulated light cover
http://www.tenmat-us.com/
Efficient Windows Collaborative
http://www.efficientwindows.org/
Nudura; Insulated Concrete Forms
http://www.nudura.com/en/nudurahome.aspx
http://nudura.com/EN/nextgen/ngvideos.aspx
http://www.youtube.com/watch?v=ycSA8l6l820&feature=related
SIPA; Structural Insulated Panel Association
http://www.sips.org/
http://www.sips.org/content/index.cfm?pageId=261
http://www.youtube.com/watch?v=eTc4wNf77rw

Lighting

U.S. Department of Energy; CFL
http://www.energystar.gov/index.cfm?c=cfls.pr_cfls
U.S. Department of Energy; LED
http://www.energystar.gov/index.cfm?c=lighting.pr_what_are
http://www.energystar.gov/index.cfm?c=ssl.pr_residential
LED Starlight Inc.; Comparison Chart
http://www.ledstarlight.com/led-comparison-chart.php
Solar Green; Tubular lighting
http://solargreen.us/products_res_daylighting.html

Microhydro

Appalachian State University
http://www.wind.appstate.edu/reports/microhydro_factsheet.pdf
Homepower Magazine
http://www.homepower.com/view/?file=HP103_pg14_New
Terratek Energy Solutions Inc. Measuring head pressure
http://terratek.ca/micro-hydro.php
Energy Bible; Weir Flow chart
http://energybible.com/water_energy/measuring_flow.html

Net Metering

Vermont Public Services; Net Metering
http://publicservice.vermont.gov/energy-efficiency/ee_netmetering.html
U.S. Department of Energy
http://www1.eere.energy.gov/solar/net_metering.html
American Wind Energy Association
http://www.awea.org/smallwind/states.html
Network for New Energy Choices
http://www.newenergychoices.org/index.php?page=nm07_NM&sd=nm

Solar

Solar Water Heating A Guide to Solar Water and Space Heating Systems From Mother Earth News By Bob Ramlow with Benjamin Nusz
Solar Energy International Photovoltaics, Design and Installation Manual
http://www.solarenergy.org/
National Renewable Energy Society
http://www.nrel.gov/learning/re_photovoltaics.html
Research Institute of Sustainable Energy, How are solar cells made
http://www.rise.org.au/nfo/Tech/pv/index.html
Sun Tracking, Solar Tracker video
http://www.youtube.com/watch?v=6j46qfuK72I&feature=related
Solar Magic, Optimizer
http://solarmagic.com/real_world
North American Board of Certified Energy Practitioners
http://www.nabcep.org/
Nanosolar "NREL Certifies 16.4% Nanosolar Foil Efficiency" Wednesday, September 09, 2009, 7:00:00 AM
http://www.nanosolar.com/company/blog/feed
Eco Geek
http://ecogeek.org/content/view/1329

Tax Credits

U.S. Department of Energy
http://www.energystar.gov/index.cfm?c=tax_credits.tx_index
http://www.energysavers.gov/financial/
http://www1.eere.energy.gov/financing/consumers.html
Database of State Incentives for Renewable & Efficiency
http://www.dsireusa.org/

Wind

Wind Energy Basics A Guide to Small and Micro Wind Systems By Paul Gipe
Wind Powering America
http://www.windpoweringamerica.gov/wind_maps.asp
http://www.windpoweringamerica.gov/pdfs/small_wind/small_wind_guide.pdf
National Renewable Energy Laboratory
http://www.nrel.gov/learning/re_wind.html
American Wind Energy Association
http://www.awea.org/smallwind/pdf/InThePublicInterest.pdf
Mother Earth News
http://www.motherearthnews.com/Renewable-Energy/2008-02-01/Wind-Power-Horizontal-and-Vertical-Axis-Wind-Turbines.aspx

<u>Notes</u>

My Powerful Home

LaVergne, TN USA
10 February 2011
216050LV00004B/275-284/P